THE SEMANTIC THEORY
OF EVOLUTION

MODELS OF SCIENTIFIC THOUGHT

A series of monographs and tracts

edited by
Roger Hahn, University of California, Berkeley.
Krzysztof Pomian, Centre National de la Recherche Scientifique, Paris.
René Thom, Institut des Hautes Etudes Scientifiques, Bures-sur-Yvette.

Volume 1

GLOSSOGENETICS The Origin and Evolution of Language
edited by Eric de Grolier

Volume 2

THE SEMANTIC THEORY OF EVOLUTION
Marcello Barbieri

Additional volumes in preparation

ISSN 0736-5268

THE SEMANTIC THEORY OF EVOLUTION

Marcello Barbieri
University of Turin, Italy

With a foreword by
René Thom
Institut des Hautes Etudes Scientifiques, Bures-sur-Yvette,
France

harwood academic publishers
chur · london · paris · new york

Harwood Academic Publishers

Poststrasse 22
7000 Chur
Switzerland

P.O. Box 197
London WC2E 9PX
England

58, rue Lhomond
75005 Paris
France

P.O. Box 786
Cooper Station
New York, New York 10276
United States of America

This book is published simultaneously in Italian under the title La theoria semantica dell'evoluzione *by Editore Paolo Boringhieri, Torini*

Library of Congress Cataloging in Publication Data

Barbieri, Marcello.
  The semantic theory of evolution.

  (Models of scientific thought, ISSN 0736-5268; v. 2)
  Bibliography: p.
  Includes indexes.
  1. Evolution. I. Title. II. Series.
  QH371.B35 1985   575   85-4756
  ISBN 3-7186-0243-1

Contents

Preface to the Series

However it is defined, the scientific enterprise is characterized by a state of permanent flux. Discoveries, inventions, languages, reformulations, and applications have succeeded one another in a bewildering variety of formats and, in modern times, with astonishing rapidity. Together with the constantly expanding domain of what is treated as scientific, the development of science and technology constantly poses problems for every thinking person attempting to understand his condition. Science's potential for change and magnification seems endless and stupefying.

This series will be a forum for discussion that penetrates beyond uncertainties and confusions. By focusing on a variety of contemporary issues connected with science and examining them from philosophical, historical, social, or other perspectives, we intend to illuminate fundamental questions that transcend the flux of scientific activity and understanding.

Models of Scientific Thought welcomes works from the whole spectrum of positive viewpoints that will help modern man reflect upon the enterprise he has invented and to fathom its character and meaning.

Foreword

Marcello Barbieri's book reads like a novel; indeed, some members of the scientific community might call this essay a 'novel', but to do so would be a mistake, for the author makes not a single claim that is not powerfully supported by facts and the logic of argument. The reader will appreciate the joyous impertinence — often tinged with real humour — with which Marcello Barbieri attacks the great taboos of current biological thought, as well as certain prejudices that persistently crop up in the evaluation of the historical merits of various biologists (on this subject, see what he has to say about Darwin and Lamarck).

The description he gives of the first stages of life, of those prebiotic cycles which led to the first cells, is as clear as the subject matter will allow. Arguably, the author's theory of the ribotype is based only on very general points. One might, for example, have wished a slightly more explicit picture of the origin of the genetic code, or an explanation of the numerical and statistical characteristics of the code but the overall aim, which is to find the origin of cellular organization in a coupling which links the polymerization of nucleic acids (dimension one) to the polymerization of proteins (dimension three), is fundamentally right: between genotype and phenotype a third element is needed, to play the role of central organizer.

The final pages, recalling as they do d'Arcy Thompson, reveal what is to my mind perhaps the biggest lacuna in the theory. Concentrating on biochemistry, which his description renders intelligible and almost enjoyable — no mean feat — Marcello Barbieri has scarcely touched upon the problems of form; problems with which our illustrious forebears of the beginning of the nineteenth century — Goethe, Geoffroy Saint-Hilaire, Cuvier, and others — were so greatly concerned. He does not mention the geometrical aspect of morphogenesis. There is nothing exceptional about this, of course; it has become customary, in biology, to consider form to be derived from chemistry and determined by it, forgetting that in basic physics quantitative laws are themselves the consequence of the geometry of space-time. In biology this geometry continues to exercise its constraints, albeit on a different scale and in a more qualitative form, one closer to topology. Perhaps the perspectives should be changed round: contingency belongs to chemistry, contingency is genetic variability, while the invariants of spatial origin remain, but are hidden ... mainly because the biologist does not have the conceptual equipment necessary to perceive them, and it is inevitable that our attention is more easily drawn to variations than to constants which escape our understanding.

I am certain, however, that this essay, thanks to the new perspectives that it offers, can do much for a better reappraisal of the problems. It is an excellent example of the new directions currently being taken in biological thinking, after decades of constraining orthodoxy imposed upon us the the central dogma and molecular biology. It should have a salutary and lasting influence, and the series 'Models of Scientific Thought' is happy to welcome it.

RENÉ THOM

Introduction

The word 'semantic' may seem a little poetic for a subject like evolution, and I would like therefore to point out that the term has recently acquired a precise scientific meaning. In biology, Emile Zuckerkandl and Linus Pauling have introduced the concept of 'semantophoretic' or 'semantic' molecules by showing that certain proteins and nucleic acids allow us to reconstruct genealogical trees and can be regarded as molecular chronometers of the history of life. In communication theory and computer science, Lila Gatlin and others have added the concept of thermodynamic meaning to that of information, a move that implies an entirely new approach to our thermodynamic description of Nature. In this way, biologists and computer scientists alike have gone beyond the stage of information only, and are beginning to speak of 'information-plus-meaning', i.e. of 'semantic' processes.

This book follows such a line of enquiry, arrives at the formulation of new principles, and shows that, taken together, these principles form a theory that gives us a new understanding of the origin and evolution of life. In the foreword, René Thom has been generous about the theory, but had to admit frankly that the book is silent on the problems of form, and of course he is absolutely right. I hope that morphogenesis and other important subjects will be dealt with in the future, and I invite the reader to expect from this book only the bare essentials of a new theoretical framework.

It is a pleasure for me to acknowledge that the title of the book was due to the suggestions of two distinguished teachers. Karl Popper pointed out that the previous title 'the Ribotype Theory of Evolution' was not entirely satisfactory even if the concept of ribotype — as you will see — is at the very centre of the new theory. Something more general was required, and it was Giuseppe Sermonti who suggested the term 'semantic' for it.

I also acknowledge with thanks the comments and encouragement I received from Jean Brachet, James Danielli, Walter Elsasser, David Horrobin, Karl Popper, Giuseppe Sermonti, Roger Stanier, René Thom, Carl Woese, and Ira Wool. And I owe a special debt to Heinz G. Wittmann and Elmar Zeitler, who supported my research at the Max-Planck-Institut für Molekulare Genetik in Berlin, where I did the experimental work which led me to this book.

MARCELLO BARBIERI

CHAPTER 1

The Idea of Evolution

THE IDEA OF CHANGE

Every schoolboy — wrote Arthur Lovejoy in 1909 — knows that Darwin did not invent the theory of evolution, and the general public is more or less acquainted with the names and the works of at least some of the earlier protagonists of the doctrine: of Darwin's grandfather Erasmus, of Lamarck, of Geoffroy St. Hilaire, and of Herbert Spencer.

This point has been emphasized by various authors ever since. In 1932, for example, J.B.S. Haldane expressed it in this way:

We must carefully distinguish between two quite different doctrines which Darwin popularized, the doctrine of evolution and that of natural selection. It is quite possible to hold the first and not the second.

It is not difficult to see why the theory of evolution should be clearly separated from the various theories that have been proposed on the *mechanism* of evolution. It is one thing to say, for example, that the sun emits light, quite a different one to say how it does it. This gives us our first problem: what is the theory of evolution when we take away from it the problem of its mechanism?

All that remains, it seems, is the idea of *change*. In 1888, Joseph LeConte summarized a widespread opinion in the following way:

Evolution is not a special theory — Lamarckian, Darwinian, Spencerian — for these are all more or less successful modes of explaining it. Evolution is a law of derivation of forms from previous forms, a law of continuity, a universal law of becoming. It is only necessary to conceive it clearly, to see that it is a necessary truth.

If this was the case, the theory of evolution would be the modern version of the old philosophies of Becoming and Eternal

1

Change. The trouble is that concepts like form, becoming, continuity and change, mean different things to different people, and the theory would be virtually meaningless. Furthermore, the idea of change has been used to support the most divergent doctrines about life. Evolution would explain everything, as LeConte said, but everything would be different to everyone, and the theory would explain nothing.

Our problem therefore becomes the following: is the theory of evolution a pack of vague generalities, or does it have an unambiguous meaning? Is it a modern theory or an old one? The best way to find out, I believe, is by looking at some ideas of the Classics. That will make us realize not only how modern and original the theory is, but also how precise its concepts really are.

"THE CHILDHOOD OF THE WORLD"

The *De Rerum Natura* of Lucretius, which appeared in 55 B.C. when the poet was probably already dead, is the most beautiful and complete synthesis of the materialist philosophy of the Classics. It describes the atomic theory, the principle that matter is neither created nor destroyed, and a breathtaking history of the Universe which is matched only by the Book of Genesis.

Multitudinous atoms, swept along in multitudinous courses through infinite time by mutual clashes and their own weight, came together in every possible way and realized everything that could be formed by their combinations. So it came about that a voyage of immense duration, in which they had experienced every variety of movement and conjunction, at last brought together by sudden encounters those which form the starting point of the substantial fabrics — earth and sea and sky and the races of living creatures.

The newly formed earth is seen as a cosmic female which was pregnant with all living organisms and Lucretius describes 'the childhood of the world' in this way.

There was a great superfluity of heat and moisture in the soil, and wombs grew up wherever a suitable spot occurred, clinging to the earth by the roots. These, when the time was ripe, were burst open by the maturation of the embryos which fled from the moisture and struggled for air. Then nature directed towards those spots the pores of the earth, making it open its veins and exude a juice resembling milk, just as nowadays every female when she has given birth is filled with sweet milk because all the flow of nourishment within her is directed into the breasts. The young were fed by the earth, clothed by the warmth and bedded by the herbage. Here then is further proof that the name of mother has rightly been bestowed on the earth.

Then, because there must be an end to such parturition, the earth ceased to bear, like a woman worn out with age. For the nature of the world as a whole is altered by age, and everything must pass through successive phases. Nothing remains for ever what it was.

The narrative continues with a description of what today we call 'the fixity of the species', 'the struggle for existence' and 'the survival of the fittest':

In those days the earth attempted also to produce a host of monsters, grotesque in build and aspect — hermaphrodites, halfway between the sexes yet cut off from either, creatures bereft of feet or dispossessed of hands, dumb, mouthless brutes, or eyeless and blind, or disabled by the adhesion of their limbs to the trunk. These and other such monstrous and misshapen births were created. But all in vein. Nature debarred them from increase. For it is evident that many contributory factors are essential to the reproduction of a species. First, it must have a food-supply. Then it must have some channel by which the procreative seeds can travel outward through the body when the limbs are relaxed. Then, in order that male and female may couple, they must have some means of interchanging their mutual delight.

The fact that there were abundant seeds of things in the earth is no indication that beasts could have been created of intermingled shapes. The growths that even now spring profusely from the soil — the varieties of herbs and cereals and lusty trees — cannot be produced in this composite fashion: each species develops according to its own kind, and they all guard their specific characters in obedience to the laws of nature.

In those days, again, many species must have died out altogether and failed to reproduce their kind. Every species that you now see drawing the breath of life has been protected and preserved from the beginning of the world either by cunning or by prowess or by speed. The surly breed of lions, for instance, in their native ferocity have been preserved by prowess, the fox by cunning and the stag by flight.

Finally, Lucretius describes the history of men, who in the beginning 'were far tougher than the men of today, and were built on a framework of bigger and solider bones because they were the offspring of tough earth'. He regards the discovery of fire, the building of huts, the development of language, the invention of agriculture and the foundation of villages and cities, as various stages in the progressive deterioration of the human condition, which gave origin to evil, injustice, fear, prejudice and eventually religion.

Again, men noticed the orderly succession of celestial phenomena and the round of the seasons and were at a loss to account for them. So they took refuge in handing over everything to the gods and making everything dependent on their whim.

Poor humanity, to saddle the gods with such responsibilities! What griefs they hatched then for themselves, what festering sores for us, what tears for our posterity! This is not piety, this oftrepeated show of bowing a veiled head before a graven image; this bustling to every altar; this kow-towing and prostration on the ground with palms outspread before the shrines of the gods; this deluging of altars with the blood of beasts; this heaping of vow on vow. True piety lies rather in the contemplation of the universe with a quiet mind.

THE IDEA OF DECLINE

The De Rerum Natura and the Book of Genesis have a few important concepts in common. All living creatures, man included, appeared on earth in a short period of creation, and came into being independently. The diversity and the complexity of life were there at the beginning, and from that time onwards the history of life was a history of deterioration and decline. We could say of devolution, not of evolution.

For Lucretius the decline was primarily biological, because the earth had become sterile and could not produce new species, except for insignificant little creatures, while many of the original species had been lost for ever. The diversity and the quality of life on earth could only decrease. The Bible does not say much on the natural history that followed the Creation, but the report that the patriarchs lived for hundreds of years (Adam lived to be 930 years old) indicates that the idea of biological decline was known and accepted. The same idea is present in the legends of many primitive tribes, which invariably speak of a 'Golden Age' at the beginning, of better, healthier and heroic ancestors.

From a religious point of view, the idea of decline makes sense, because God is perfect and must have created a perfect world. But He also allowed freedom to exist, and therefore evil, and evil could only lead to a deterioration in the original word.

What is less obvious is why the idea of decline was shared even by men like Leucippus, Democritus and Lucretius who professed a total materialism. Why did they think that life had to explode at the beginning in the full glory of its innumerable forms? Why did not they, or their followers, conceive the idea that only a few simple creatures appeared on earth, and slowly evolved into more diverse and more complex forms?

After all, the universe had originated, for them, from the chance encounters of atoms which arranged themselves in all possible combinations. There had been therefore an increase in complexity in the physical world, from the lowest to the highest possible levels. Why did they not extend this idea to the biological world?

There may be many good reasons for this, but one, I believe, is fundamental. They conceived the idea that the physical world is made of units, or atoms, but nobody thought that the biological world is also made of units, or cells. If they had thought of that, it would have been possible to imagine that the cells gave origin to increasingly more complex combinations as the atoms did, and the idea of biological evolution could have been born.

But they didn't. The classic philosophers conceived many ideas that later became science; they thought not only of atoms, energy and wave motion, but also of relativity and indeterminism. For some reason, however, the idea of the cell was unthinkable, and without it the concept of organic evolution could not germinate.

Some people may think that this explanation is too trivial, but they should think again. We are far too accustomed to take the cell for granted, and are largely unaware that to think of life in terms of cells has been one of the greatest theoretical revolutions in the history of mankind.

At any rate, it is an historical fact that the idea of evolution from microorganisms to animals and plants started circulating *after* the discovery of the microscope and of miscroscopic life. That, in my opinion, is not a coincidence. Just as it is not a coincidence that our understanding of the cell and of evolution have gone hand in hand ever since.

A NEW BEGINNING

In the previous section we have already encountered the most important idea of the theory of evolution: *life appeared on earth in the form of one or a few types of microorganisms.*

As we have seen, both the De Rerum Natura and the Bible spoke of a link between present and past organisms; the idea that life had a history and that changes occurred during this history was already there, and there is nothing new in it. The real difference between Evolution and all previous concepts of historical Change is the idea of what happened in the beginning, of what creatures came into existence first.

If only a few types of organisms appeared on the primitive earth, then a great process of diversification must have taken place during the history of life. If the first creatures were invisible microorganisms, their descendants had not only to become different, but some of them had to develop into bigger and more complex forms of life.

Evolution is the idea of an historical change accompanied by increasing diversity and increasing complexity, and its very foundation, therefore, is the hypothesis of the ancestral microorganisms, the idea that life on earth began with one or a few types of cells.

Today we look at this hypothesis from a rather comfortable position. We have a fairly good idea of the age of the earth — around 4.6 billion years — and from the fossil record we know that for almost three thousand million years — roughly from 3.6 to 0.6 billion years ago — the microorganisms were the sole inhabitants of our planet. They were not only the first creatures to appear, but had the earth to themselves for about two thirds of its history, and today they are still the predominant form of life (they account for at least 95% of all living matter).

All this information however is recent history. The first microfossils were discovered in 1954, and before that date there wasn't the slightest evidence for the idea of ancestral microorganisms. How was it possible then to conceive such an idea in the 18th century? How was it that a few men of science came to think what no philosopher had ever thought before, what no poet had imagined and no experimental fact had documented? The story began, as so many others in science, with the invention of an instrument.

Antoon van Leeuwenhoek announced the discovery of the microscope in a letter to the Royal Society of London in 1676, and in 1683 sent another letter that described the first 'microscopic observations about animals'. His drawings, published in

1684, were the very first pictures of bacteria and protozoa. Leeuwenhoek was also the first person to observe spermatozoa under the microscope, and it was this report, more than anything else, which fired people's imagination and spread enthusiasm for the new science of microscopy.

Here at last one could *see* what is starting life: a tiny, wriggling, tadpole-like infinitesimal creature which had a round head and a tail. The excitement produced a chaotic stream of reports, and it is difficult for us to differentiate what people were seeing under the microscope from what they imagined they were seeing, but that is understandable. Scientists had to get used to the idea of microscopic life, had to convince themselves that microorganisms are not freaks, that they are everywhere and not just in some curious places, and that some progress could be made by building better microscopes.

Microbiology and Embryology grew very slowly, but they grew together, and inspired a diffused belief in the idea that there is a parallel between the general development of life and the individual development of the embryo. In 1744, Albrecht von Haller coined the word *evolution* to indicate the unfolding of an embryo, but later the word came to be used for the development of life as a whole. This linguistic transformation reveals only too clearly the emerging feeling of the time. If the beginning of every organism is a single cell, perhaps life on earth started in the same way, from invisible microscopic creatures.

Today the origin of the cell is an enormous puzzle for us, but in the 18th century it was no problem at all. The belief in spontaneous generation was genuine and widespread (despite the experiments of Redi and Spallanzani), and provided the perfect answer. All the necessary ingredients therefore came together. Microscopic life is full life, it can produce visible creatures, and can originate naturally and suddenly by spontaneous generation. The idea of evolution 'in principle' was born.

The fact that this was pure speculation should not deceive us. Since there was no evidence, scientists pushed the idea of the ancestral microorganisms into the background, and concentrated on the other problems of the history of life, in particular on the 'transformation' of species. But that 'speculation' was the very heart of the young theory, because there would be no point in talking about an increase in diversity and complexity if the original creatures had not been less diverse and less complex.

The microscope made it possible to see and think what nobody had seen and thought before, and it doesn't really matter if scientists did not talk or write openly about life beginning from invisible creatures. The important point is that for the first time in history they could think about it and find it *credible*. The formal publication of the new theory, at any rate, was not far away.

LAMARCK

The book which put together the speculations of the 18th century into the first complete theory of evolution, was published in 1809 (the year in which Darwin was born). It was entitled 'Philosophie zoologique', by Jean Pierre Antoine Baptiste de Monet, le Chevalier de Lamarck, and the theory consisted of three hypotheses:

1) Life appeared on earth in the form of microorganisms, by spontaneous generation.

2) Organisms diversified and adapted to the environment by the inheritance of acquired characteristics.

3) Organisms became more complex because there is an intrinsic tendency in every creature to reach higher levels of complexity.

It is no secret that in every one of these hypotheses there is something wrong. The idea that microorganisms can originate by spontaneous generation was definitely abandoned after the experiments performed by Louis Pasteur in 1860. The idea that acquired characteristics can be inherited survived a bit longer, but eventually became incompatible with too many results in Genetics. Finally, the idea that organisms have an intrinsic tendency towards complexity is too vague and obscure to be of any use; without a more specific definition, it amounts to saying that organisms become more complex because they tend to become more complex, which explains nothing.

Lamarck, on the other hand, saw in this obscure force the effective driving power which shaped the history of life. 'Evolution by Internal Drive' was the mechanism proposed by Lamarck as 'Evolution by Natural Selection' was the mechanism proposed by Darwin.

There is however another way of looking at Lamarck's contribution. He proposed the wrong mechanism, but we should keep in mind that the theory and the mechanism of evolution must be clearly separated. This can be done easily by eliminating the last part from Lamarck's hypotheses, which now read as follows:

1) Life appeared on earth in the form of microorganisms.

2) Organisms diversified and adapted to the environment.

3) Organisms became more complex.

These are the three basic hypotheses of the theory of evolution: the origin of the cell, the origin of diversity and the origin of complexity, all by natural causes. Lamarck was truly the first person who formulated the theory correctly.

There is another fact which proves this point. The visible expression of the theory of evolution is the 'Tree of Life', a diagram which represents different creatures linked together as if they were the members of the same genealogical tree. Lamarck was the first person to draw such a diagram. It was a very simple sketch (Figure 1), with only a few entries, but the essence of what we call phylogeny (the development of the species) is all there.

Again, the details are wrong (he proposed that whales and land mammals derived from seals, for example), but that is hardly surprising. Darwin did not even attempt to draw a real phylogenetic tree (the diagram that he put in the Origin of Species — reproduced in Figure 2 — does not represent actual organisms), and the first complete phylogenetic tree, published by Ernst Haeckel in 1866 — does not have fewer mistakes than Lamarck's (Figure 3).

The first orang-utan that was brought to Paris caused a sensation, according to Diderot, and in 1770 De Lisle de Sales suggested that the orang-utan had been the ancestor of man, for which he was duly sent to prison. The idea that we descend from the apes was interesting or disturbing, according to one's point of view, and in any case dangerous, but it had not required a great leap of the imagination: a good look at an ape had been enough. It is one thing, however, to recognize familiar features in apes,

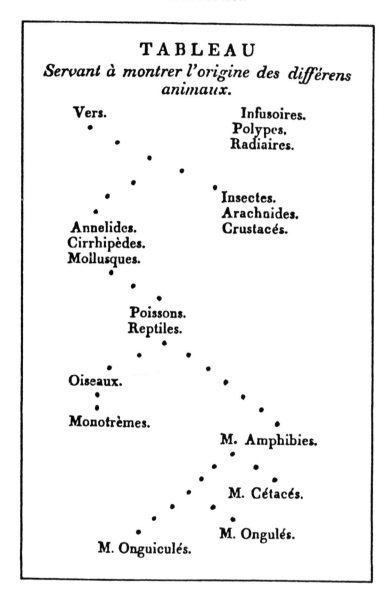

FIGURE 1:   The first phylogeny of animals, published by Lamarck in 1809.

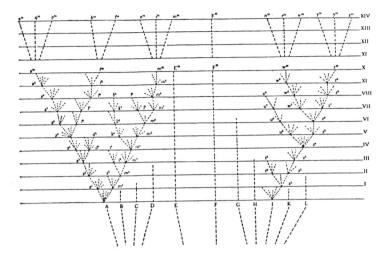

FIGURE 2: Darwin's illustration of the principle of divergence and the origin of species (1859).

quite another to say that we are related to birds, snakes, whales and many other animals around us. That is truly beyond all appearances.

It is odd that people have been unable to give Lamarck credit for this great vision, and remember him only for the inheritance of acquired characteristics. Darwin too proposed a theory of inheritance which is equally discredited (the theory of Pangenesis), but this is hardly mentioned. Why people chose to remember Darwin only for his good ideas and Lamarck only for his bad ones is a mystery to me.

At any rate, when Lamarck published the first phylogenetic tree, in 1809, he effectively wrote the 'manifesto' of evolution, the programme for an immense research project which is still going on. The theory of evolution was born then and there. The great achievement of Darwin, 50 years later, was to find a convincing mechanism for it, and make it popular.

NATURAL SELECTION

Did Darwin solve the problem of the origin of species? Has the

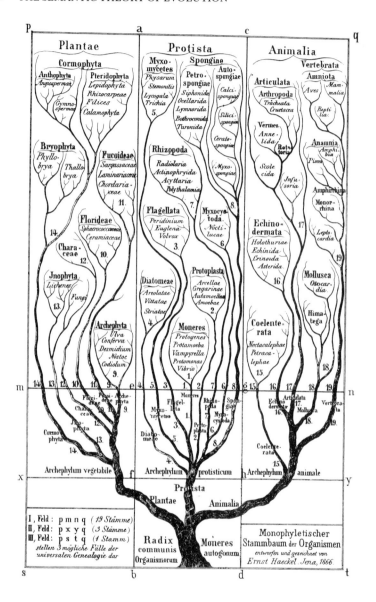

FIGURE 3:   The first phylogenetic tree from bacteria to man, published by Haeckel in 1866.

problem been solved since? To these questions some biologists reply yes and others no, which means that the problem is still open and worth discussing. To start with, let us examine a few examples, with the reminder that individuals, in Biology, are grouped in varieties, varieties in species, species in genera, families, orders, classes, phyla and kingdoms.

When the Penicillium mold was first isolated, its natural production of penicillin was 30 units per milliliter, and in order to increase this yield the pharmaceutical companies embarked on one of the largest programmes of biological modification that has ever been undertaken. Today, the industrial descendants produce 30,000 units of penicillin per milliliter, and yet for all the variations and selections that it has gone through the mold is still a Penicillium, it has not changed species (Sermonti, 1981).

The same is true for the experiments where a preferential outcome was deliberately avoided, and biologists just watched what happened in cell cultures under different conditions. The appearance of new varieties of microorganisms from a single strain has been described in detail countless times, but not the appearance of new species.

Let us examine these experiments with cell cultures a little more closely, because they are an invaluable source of information: we can easily grow billions of cells in small dishes, and observe many generations of descendants in a relatively short time.

If we divide a heterogeneous population of cells in equal groups, for example, and cultivate them in dishes which provide different environments, some generations later we usually find that each dish has its own predominant form of microorganisms. We learn in this way that the strength or the weakness of each form is not an absolute quality, but something that depends very much upon the environment.

Let us now take groups of identical cells and grow them in conditions which are kept rigorously equal in all dishes, but which are not ideal for that type of cell. Some generations later, we usually find that in some dishes the cells have changed and grow better than in others. The important point here is that at the beginning of the experiment there was no way of deciding in which dish the changes would take place, when and if they would take place, and what types of change would take place.

We learn in this way that biological variations occur at random.

In these experiments we come face to face with two basic processes. We see that changes in hereditary characteristics take place at random, and we see that the environment has a critical role in determining which variety has more success in reproducing itself. As if the environment had an invisible hand which operates a natural process of selection.

Variation and selection are clearly different things, but biologists have been used to describing them as the two phases of one process, and have given to this two-step mechanism the name of "natural selection". This is not the clearest way of describing what happens, but the terminology doesn't really matter. The important point is that we can convince ourselves that natural selection not only exists, but really works: we can actually see it in action before our own eyes in the laboratory.

But we also see that only new varieties appear, not new species, as if species had the ability to bend but not to break when the environment changes. Natural selection, in other words, has all the characteristics of a mechanism that works for the conservation of existing species, not for the creation of new ones.

This conclusion is by no means valid only for microorganisms. Centuries of crossing and breeding experiments with animals, crops and plants, sometimes performed on massive scales, have shown the same pattern time and time again. Many new varieties have been obtained, but no human being has ever seen and recorded the formation of a new species in Nature.

DARWIN

According to Loren Eiseley (1959), the earliest mention of "a natural process of selection" appeared in 1831 in a book by Patrick Matthews, but the first interpretation of its role in Nature was given in 1835 by Edward Blith, who described it as a strongly *conservative* force. In 1837, Blith asked himself if the cumulative effects of natural selection could have led in the long run to the creation of new species, and answered in the negative: to him, biological adaptations have the net result of conserving existing species, not of creating new ones.

As we know, 21 years later Charles Darwin and Alfred Russel Wallace examined the same question in their memoirs to the Linnean Society (read together on July 1st, 1858), and answered in the affirmative: natural selection is *the* mechanism by which new species are formed during the history of life.

Did Darwin and Wallace have new conclusive evidence? No, they didn't have a single case in which the origin of a new species had been seen and documented. Did they show that Blith's reasoning was wrong? Again, no. The conclusion that species conserve themselves by adapting to the environment was (and still is) based on too many facts to be challenged. How then did they manage to turn the role of natural selection upside down, and convince quite a number of biologists that that most conservative force is in fact the creative power of evolution?

Darwin explained this in detail in the *Origin of Species* (1859). What the evidence shows, he argued, is that variations do occur in the hereditary characters, and that they occur at random. Furthermore, the evidence shows that variations usually come in *small jumps*, because monstrosities or sports as a rule are not viable. There is therefore in Nature what Wallace called *the Tendency of Varieties to depart indefinitely from the Original Type*, and which Darwin described as *the Principle of Divergence*.

Let us now attribute this natural property of divergence to the most primitive creatures, and follow with the eyes of the mind what happened. The descendants of the ancestral creatures gave origin to new varieties, the descendants of each variety diversified further, and the descendants of the descendants went on diverging in countless varieties, like a tree that started from one seed and keeps producing more and more branches.

But not all varieties could survive. Many died, and left a complicated series of gaps in the network. In this way, the branches of the Tree of Life formed irregular patterns, like those of a real tree, and arranged themselves in major, medium, minor and minute groups to which we give such different names as phyla, orders, genera and species.

When we look at different species we may think that a creative force had to be responsible for the diversity, while in fact it is only because their intermediate forms disappeared that we perceive them as separate groups. The discontinuity between species is real only because the gaps were produced by real extinctions, not because there was a creative process of *speciation* which differs in principle from the standard process that gives origin to the varieties.

Ultimately there are only varieties in Nature, according to Darwin:

Varieties, then, have the same general character as species, for they cannot be distinguished from species ... except ... by a certain amount of difference, for two forms, if differing very little are generally ranked as varieties, notwithstanding that intermediate linking forms have not been discovered; but the amount of difference considered necessary to give to two forms the rank of species is quite indefinite.

That was Darwin's solution. There are no longer barriers between species, only gaps, which means that they are all related. The "mystery of the mysteries" simply disappears, and in its place we are given the unity of life.

Lamarck, as we have seen, had already spoken of links between different species, but he had also believed in spontaneous generation, and therefore in the possibility that different microorganisms appeared at different times and became the progenitors of separate trees of life. In Darwin's framework, spontaneous generation has no place, and the unity of life therefore runs all the way through the organic world, embracing all creatures of the present and linking them to all creatures of the past.

We are here today because there came before us "a finely graduated organic chain", a procession of "a truly enormous number of extinct varieties" which have slowly built all the steps of all the routes that link every creature to the original ancestor.

Darwin summarized this vision in the final words of his book:

There is grandeur in this view of life, with its several powers, having been originally breathed by the Creator into a few forms or into one; and that, whilst this planet has gone cycling on according to the fixed law of gravity, from so simple a beginning endless forms most beautiful and most wonderful have been, and are being, evolved.

CUVIER

Various fossils had been collected throughout the 18th century and attracted a certain amount of interest, but the interpretations were vague and often extravagant, and Paleontology remained little more than a curiosity until the end of the century. What transformed it into a science were the works published by George Cuvier between 1798 and 1812.

Cuvier held the chair of Anatomy at the Museum National d'Histoire Naturelle in Paris (the Institute where Lamarck was working), and it was with precise anatomical arguments — he was the founder of Comparative Anatomy — that he gave substance, method and credibility to the study of fossils.

The first problem we face in Paleontology is that, in most cases, all we have of past organisms is a few bones, and these would tell us very little if it were not for the first golden rule of Cuvier, the *Principle of the Correlation of Parts.* This principle tells us that we can reconstruct a whole body from a small portion of it because every part is correlated with the others and in general is compatible with only one anatomical architecture. Cuvier himself made the drawings of entire organisms from a few bones, and when complete skeletons were discovered, after his death, people were amazed to find how close he had come to the real structures, even in small details.

The reconstruction of a whole body from a few remains, however, is only the first step. After that we have to place the reconstruction in its historical context, and here too we follow Cuvier. Let us illustrate this with an example. There were at the Museum bones of elephant-like creatures which had been found in Siberia, and in 1798 Cuvier was able to demonstrate that they had belonged to a species which differed from all existing elephants, the extinct species of mammoths.

It was the very first demonstration that species which once lived on earth had become extinct. More important, the demonstration was based on anatomical principles which are valid in general, and not just for that particular case, and in this way Cuvier gave us the second golden rule of Paleontology. Once we have reconstructed the structure of a past organism, we establish with Comparative Anatomy its relationship with known forms and decide if the creature in question has become extinct or not.

In addition to this theoretical work, Cuvier went in person to excavate around Paris, and made another fundamental discovery. He found that the fossils in the chalk caves of Montmartre were distinctly different from those of the strata immediately above and below the caves. One layer had fossils of creatures which lived in the sea, the second of fresh-water organisms, and the third of animals which lived on dry land.

A succession of geological layers was revealing a succession of periods in the history of life, as if the layers were the pages of a giant book. It became possible to establish which creatures came before which, and from the thickness of the layers one could estimate, approximately, how long each period had lasted in respect to the others. Cuvier had discovered Stratigraphy, the most important tool of the paleontologists.

As for the theoretical implications, Cuvier was well aware of the ideas of Lamarck and his fellow 'transformists', of the suggestion that species are not immutable, that organisms had undergone gradual and progressive changes during the history of life, but he did not believe them. It may seem strange that the man who created Comparative Anatomy, Paleontology and Stratigraphy could not accept the idea of a gradual evolution of life, but he had two reasons for this.

In the first place, he firmly believed in the supreme priority of the facts, and to him the facts of the fossil record were saying that species had not changed gradually. They had existed unmodified for long periods of time, and then disappeared suddenly from the record, only to be replaced with equal suddenness by new species. The geological periods succeeded one another abruptly — one could see this clearly written in the rocks — and the same was true for their forms of life: there was no trace of intermediate organisms in the record.

Cuvier candidly confessed that to him the sudden appearance of new species was the greatest mystery of all, but the records said that that is what happened, and to him that was that. The history of life had been a history of catastrophic extinctions followed by explosions of new forms, and no speculation could change that hard fact of Nature.

The second reason for rejecting evolution came from his discovery of a new biological category that he called 'embranchement', and that today we call 'phylum'. He compared various species of animals and discovered that the immense variety of body-structures that we find in Nature can be reduced to a few basic anatomical designs.

In the wings of birds and bats, for example, we find the same types of bones that exist in our arms, in the flippers of seals and in the forelimbs of reptiles and amphibians. The size and the shape of each bone varies enormously from one species to another, and yet we can clearly recognize which is which simply by its position and its relationship to the other bones. We say that the forelimbs of all vertebrates are *homologous*, because they have the same structural design, as if they had been obtained by modifying a common prototype-limb.

The wings of insects, on the other hand, have no bones. They are *analogous* to the wings of birds because they are used for the same function, but their structural plan is totally different. The important point is that a structural difference in a single part means that the anatomical plan of the whole body is different.

All animals which have a skeleton inside the body, for example, have the nervous system in a dorsal position and the heart in a ventral position. Those which have an outside skeleton (insects, crustacea and spiders) have that order reversed: the nervous system is ventral and the heart is dorsal.

All organisms that have the same basic anatomical design form one group — a phylum —, and Cuvier proved that there are at least 4 different phyla in Nature. Today most biologists recognize at least 25 phyla, and it is universally accepted that the phylum is a biological category as fundamental as the species. The anatomical barrier between different phyla is as real and as natural as the reproductive barrier between different species.

The existence of separate phyla was for Cuvier incompatible with evolution. In principle he could concede (if it wasn't for the evidence of the fossil record) that a prototype-limb had the potential to become a hand, a flipper or a wing (the homologies are only too obvious), but a common origin for the wings of insects and birds, for example, was anatomically absurd.

Transformation within a phylum are conceivable, but transformations from one phylum to another are impossible, and since the phyla are irreducible, concluded Cuvier, they cannot have a common origin.

In an historic public debate with Geoffroy St. Hilaire, on February 15th 1830, Cuvier formulated his two great objections to evolution: the discontinuity of the fossil record, and the discontinuity of the biological phyla. They were directed against the theory of Lamarck (who died in 1829, blind and destitute, at the age of 85), but since Darwin's theory is equally based on gradualism and continuity, these objections have been the two most powerful arguments of the antievolutionists ever since.

BOREDOM AND TERROR

In the 1830s, Roderick Murchison made what is probably the greatest paleontological discovery of all times. He found that the rocks of the Cambrian period contain fossils of all the major animal phyla, while the strata underneath them have no visible trace of life. The appearance of many different types of complex animals in the fossil record is so abrupt that it has become known ever since as "the Cambrian explosion of life".

To Murchison, the discovery meant that he was witnessing nothing less than Creation itself: the Cambrian had to be the period chosen by God to populate the planet. Today, the precambrian rocks have been examined under the microscope and the fossil remains of microorganisms have been found in many of them, but Murchison's discovery has not lost its impact. The gap between precambrian forms and animals already differentiated into various phyla still represents a jump of immense proportions.

The Cambrian explosion is the most remarkable discontinuity in the record, but at least four other major upheavals took place afterwards, at the end of the Devonian, Permian, Triassic and Cretaceous periods. They consisted of mass extinctions followed by the rapid expansion of new forms of life.

The worse catastrophe took place at the end of the Permian when, according to the catalogue of David Raup (1979) up to 96% of all species were destroyed "leaving as few as two thousand forms to propagate life" (Gould, 1981). The catastrophe was so indiscriminate that, according to Stephen Jay Gould, it would be more appropriate to talk of the "survival of the luckiest" instead of the "survival of the fittest" for the organisms that remained alive.

The most popular catastrophe, the one which wiped out the dinosaurs at the end of the Cretaceous, also cut across many barriers and hit indiscriminately: only small animals, weighing no more than 20 pounds survived (Valentine, 1978).

The evidence of almost two hundred years of Paleontology leaves no doubt that the pattern of abrupt change described by Cuvier is substantially correct, and is valid not only for a few great episodes, but represents the rule, rather than the exception, throughout the history of life.

Instead of finding the gradual unfolding of life, what geologists actually find is a highly uneven or jerky record; that is, species appear in the sequence very suddenly, show little or no change during their existence in the record, then abruptly go out of the record.

These words, written by David Raup in 1979, are essentially what Cuvier had said 150 years earlier.

The theory of *Punctuated Equilibria*, proposed by Niles Eldredge and Stephen Jay Gould in 1972, describes the same scenario: the history of life has been a sequence of "punctuations" or rapid changes, followed by "equilibria" or long periods of stability. "It was — wrote Derek Ager — like the life of a soldier ... long periods of boredom and short periods of terror".

There is however a fundamental difference between our perception of the past and that of Cuvier: radioactivity has allowed us to measure the absolute age of the rocks, and has enormously extended the temporal dimension of geology. Now we know, for example, that the *peak* of the Cambrian explosion (some 600 million years ago) lasted approximately 50 million years, while its *tails* covered between 150 and 200 million years. The dinosaurs disappeared some 65 million years ago at the end of a process that lasted at least 10 million years, and the same amount of time elapsed in the Great Catastrophe of the Permian, around 225 million years ago.

An interval of 10 million years is less than 0.3% of the age of the earth and in geological terms is short, but in biological terms — in terms of numbers of generations — it is a very substantial length of time. What was sudden for Cuvier is still geologically sudden for us, except that instead of centuries we are now talking of millions of years, and this does make a difference. An abrupt transition in the fossil record may look as if the change took place from one generation to the next, while in fact there was time for hundreds of thousands of generations even in the fastest transitions on record.

This leaves the door open to the *possibility* that new forms of life derived from previous forms in rapid bursts of evolutionary activity, and recently the fossil record has revealed a pattern which is consistent with this interpretation.

MICROEVOLUTION AND MACROEVOLUTION

The Cambrian explosion has been intensely reinvestigated in recent years, and in 1978 John Sepkoski published a mathematical analysis of the phenomenon that casts a new light on it. He divided the Cambrian period and those which came immediately before and after it into sub-periods, counted the species which are known for each of them, and plotted the results on a diagram. The curve that came out has the typical shape which describes the growth of a population: an initial period of slow increase is followed by a rapid expansion and then by a stationary phase where the colony as a whole stops growing.

The diagram of Sepkoski and other similar models (Stanley, 1975 and 1979; Gould, Raup, Sepkoski, Schops and Simberloff, 1977; Gould, 1978) gave new life to the old analogy that a species is not just a collection of individuals but a super-individual, a natural organism made of distinct creatures as any creature is made of distinct cells. According to the analogy, species are born, live and die as individuals do, for example as the microorganisms of a cell culture. The analogue of a cell is a species; the analogue of cell division is *speciation*; the analogue of cell death is the *extinction* of a whole species.

One may point out that the analogy requires a very high number of past species to be valid, but this is not an obstacle. At least 99.9% of the species which appeared since the Cambrian explosion have become extinct, and since there are today at least 4 million species on earth, this puts the total number of past species at about 4 thousand millions, at a conservative estimate.

If the analogy is valid, the Cambrian explosion was the rapid phase of growth of a colony of species, no more mysterious than the rapid phase of growth of a colony of cells, and the phenomenon therefore becomes a perfectly natural event. The analogy can then be taken a step further. If we damage a cell culture, or subtract food, the cells start dying and the colony faces extinction; if we restore viable conditions in time, the colony expands again. In this way we observe a pattern of extinctions, expansions and long periods of stability that gives us a model not only for the Cambrian explosion, but also for what happened throughout the history of life.

But there is a price to pay: the analogy stands only if we admit that speciation is as real as cell division is, and at present we know nothing about the genetics or the biochemistry of speciation. New species may have originated by the accumulation of many small jumps — as Darwin suggested — or by a totally different mechanism: we simply do not know.

Models for an alternative mechanism of speciation have been advocated ever since Darwin's times, and are still actively debated. In 1940, for example, Richard Goldschmidt gave the name of *micromutations* to the genetic changes that produce new varieties within a species, and *macromutations* to those that produce new species altogether (by "a repatterning of the chromosomes"). There are two distinct types of genetic change, according to Goldschmidt, and therefore two distinct types of historical change which he called *microevolution* and *macroevolution.*

Microevolution does not lead beyond the confines of the species, and the typical products of microevolution, the geographic races, are not incipient species. There is no such category as incipient species. Species and the higher categories originate in single macroevolutionary steps as completely new genetic systems (Goldschmidt, 1940).

Today, the idea of macroevolution is supported by an increasing number of biologists, and in particular by paleontologists.

The reductionist view that evolution can ultimately be understood in terms of genetics and molecular biology is clearly in error. We must turn not to population genetic studies of established species, but to studies of speciation and extinction in order to decipher the higher-level process that governs the general course of evolution (Steven Stanley, 1975).

Perhaps this is true, but the fact remains that speciation is still a biochemical mystery. What has dramatically changed, however, is the framework in which we perceive the mystery. Darwin argued that intermediate varieties must have existed between the species, and arrived in this way at the conclusion that all creatures are related. Today we are not at all sure about the intermediate varieties, but we are positive about an underlying network of relationships in all organisms. The unity of life has a solid biochemical basis in the fact that the genetic code is universal, and that all cells use ATP (Adenosine triphosphate) for energy and share a wide variety of metabolic mechanisms.

What was a conclusion for Darwin, is therefore a starting point for us. The beginning of a process that hopefully will reconcile paleontology with evolution, or Cuvier with Lamarck and Darwin, after more than a century of bitter controversies.

TIMETABLES AND PRIORITIES

The earth is 4.6 billion years old; single cells were already on it nearly 3.6 billion years ago, and multicellular life exploded about 0.6 billion years ago. These dates divide the history of life on earth into three great chapters: 1000 million years of mystery; 3000 million years of microorganisms, and 600 million years of visible life. This last period — which paleontologists call *Phanerozoic* (the age of manifest life) — represents 13% of the story, it is only the final chapter of it.

In the 19th century, the period that came before was regarded as a mere preliminary stage, and was simply called *Precambrian*; now we know that that period covered 87% of the history of the earth. The feeling that we have been studying only the last episode of a very long story, that perhaps we should rearrange our approach and our priorities, is only too natural in these circumstances.

There are some, however, who resist this change on the grounds that time is not everything. It is possible, for example, that during the 3 billion years of cellular history life developed extremely slowly. Nobody denies that that period was important, but probably it was not as important as its enormous duration seems to suggest. Perhaps things really started to happen fast only with the Cambrian explosion, and what came before was only an immensely slow and sluggish preparatory stage.

This point of view has some justification if one looks at appearances: even in the Cambrian, for example, life was very primitive indeed. No animal was walking on the land, no insect nor bird was flying, no grass, no plants and no flowers were decorating the landscapes, and probably no fish were swimming in the seas.

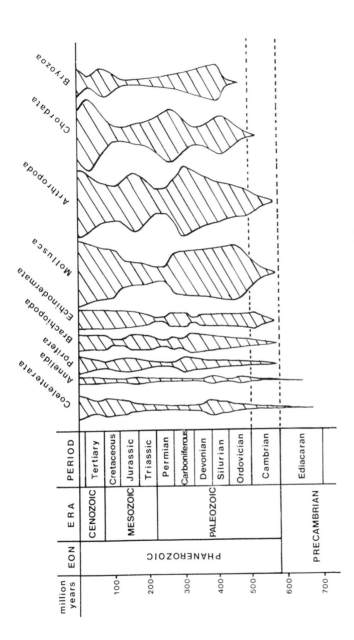

FIGURE 4:   Variations of the major animal phyla since the Precambrian (redrawn from *The Meaning of Evolution* of George G. Simpson, 1949).

And yet, crawling at the bottom of the oceans, there were strange-looking creatures which had already developed the characteristics of all major animal phyla. We must look at them "with speculative eyes", said George Gaylord Simpson, otherwise we understand nothing because external appearances are to a large extent irrelevant in Biology.

Simpson summarized the history of the Phanerozoic with the scheme of Figure 4, which shows that no new major phylum has been invested since the Cambrian explosion. Furthermore,

in spite of possible exceptions involved in the largely verbal question of defining 'phylum', it remains true that no major basic type of animal organization is known ever to have become extinct (Simpson, 1949)

From the Cambrian onwards, in other words, nothing has been created or destroyed at the level of the phyla and above. All that happened were transformations within the phyla, rearrangements of the same anatomical designs, variations on the same basic themes, often spectacular, but always contained within the potentials of the existing phyla.

The invention of countless variations from a few designs has changed the face of the earth many times over but cannot be compared, in quality, with the invention of the basic designs. As Simpson put it:

The Cambrian age of life already represents an increase which throws into shade the increase from Cambrian to Recent.

This is why the history of the Precambrian is so overwhelmingly important: not because it lasted so long, but because the fundamental developments took place at that time. The origin of the cell in the early Precambrian, and the origin of multicellularity, at the end of it, were the two main events of evolution.

Three centuries ago, the microscope revealed a new world of life, and the story of evolution began when people started *imagining* ancestral microorganisms. Since 1954, the ever-surprising microscope has shown that that imagined world is an historical reality — a three billion years long reality — and evolution has taken a dramatic new turn: we are going back to the microorganisms. After so much wandering, we are at last beginning at the beginning.

CHAPTER 2

The Precambrian

THE 'STONY CARPETS'

At the beginning of our century, Charles Walcott discovered a new type of rocks in North America. He found mounds, domes and pillars of various dimensions and shapes which had unique characteristics: they were made of wafer-thin layers, and were definitely precambrian. The layers formed flat or curly mats, sometimes circular ones, but in all cases their pattern had, like the rings in trees, that irregularity of execution which is often a sign of biological activity.

Walcott suggested just that: as whole islands have been built by corals, those domes and pillars had been left behind by precambrian microorganisms. The cells had gone, but the minerals deposited on them had solidified and had become the stony replicas of the colonies. They were the empty fossilized cities of precambrian empires.

A few enthusiasts supported Walcott's interpretation, and in 1908 Kalkowsky gave these formations their present name of *stromatolites* (stony carpets). Most geologists however were sceptical and diagnosed a non-biological origin for them. The search for precambrian life went on for many more years amid failures, false alarms and general disbelief.

The breakthrough came in 1954, when Stanley Tyler and Elso Barghoorn examined thin slices of rocks from the Gunflint deposits, near Lake Ontario, under the microscope, and found a jungle of 2-billion year old microfossils. In this case, the cells had fallen into a solution of colloidal silica, or had been embedded in a limestone that was later replaced by silica, and in time the embedding matrix had crystallized into a very hard rock called chert. In a few cases the most delicate details had been preserved, to the point that some microfossils could be classified as distinct species. The discovery launched a world-wide hunt, and a few other 'gold mines' were found. Then reports came that in some inhospitable places (for example at Shark Bay in Australia) there are stromatolites which still have living inhabitants, and whose microorganisms are strikingly similar to some of the types which had been buried two or three billion years earlier.

The picture of the Precambrian which has emerged from these discoveries is still very fragmentary, but some general features, as we will see, can be recognized and a few valuable conclusions can be drawn. The oldest terrestrial rocks come from the Isua mountains, in Greenland; they are 3.8 billion years old, and do not appear to contain fossilized cells. The oldest microfossils, so far, have been discovered in the Warrawoona formation, in Western Australia, and are approximately 3.4 billion years old (Awramik, 1980). These dates do not come anywhere near a satisfactory dating of the origin of life, but they are all that we have, at the moment. While keeping an open mind, therefore, our present estimate for the appearance of the first cells can be set somewhere in between the above figures: around 3.6 billion years ago.

In this way, we can divide the Precambrian into two great phases. The first billion years — from 4.6 to 3.6 billion years ago — was apparently without cells, and can be referred to as *Prezoic* (before life) or *Eozoic* (the dawn of life). The remaining three billion years were cell history, and we may call them collectively *the Age of the microorganisms*.

THE IRON BANDS

After the excitement of the first discoveries, scientists had to face the problem of making a realistic assessment of precambrian life. Did the microorganisms fill the oceans to capacity or were they condemned to a meagre existence, like their descendants in Shark Bay? Was the Precambrian a Garden of Eden for them, or a desert with only a few veins of life?

In the great majority of precambrian rocks there are no signs of life, this is a fact. One could say that the signs have been erased later, when the rocks were greatly transformed by heat and pressure, but this is only an hypothesis. From their lack of fossils one could also conclude that most rocks had never contained living forms, and therefore that life had been extremely scarce in precambrian times.

Luckily, there is a way of settling this problem. All over the world there are rocks from the middle Precambrian which consist of alternating layers of iron-rich and iron-poor silica (Banded Iron Formations), and these represent a planetary transition between two distinct phases in the geology of iron. In the absence of oxygen, iron is deposited in ferrous salts which are relatively soluble in water, while in aerobic conditions it produces salts of ferric iron which are insoluble. The sediments of the Iron Bands indicate therefore that a world-wide transition from ferrous to ferric salts was taking place, in other words that oxygen was accumulating in the atmosphere.

The process lasted nearly a thousand million years, roughly from 2.6 to 1.6 billion years ago, with a peak between 2.3 and 1.9 billion years ago which corresponds to the bulk of the Iron Bands. The most important point, for us, is that the oxygen which changed the earth's atmosphere could only have been released by microorganisms. It had to be then, as it is today, the waste-product of photosynthesis because there is no other natural phenomenon that could have caused a release of that magnitude.

The fossil remains of photosynthetic microorganisms, and in particular of bluebacteria (also known as cyanobacteria or blue-green-algae), have been found in precambrian rocks, and there is no doubt therefore that they existed at that time, but we can conclude more than that. The transformation of the whole atmosphere, as well as of the oceans of the planet, was an operation of immense proportions, and could not have taken place if life was barely surviving in a few isolated areas.

We conclude therefore that in the middle Precambrian the oceans were swarming with armies of microorganisms, and in particular with bluebacteria, which are the mot numerous forms in the precambrian record and also the most versatile photosynthetizers. More precisely, we can say that the eon between 2.6 and 1.6 billion years ago was the age of the oxygen revolution on earth. We can rightly call it *Cyanozoic*: the Age of the Bluebacteria.

THE POISON OF LIFE

The evidence of the Iron Bands gives us a precious piece of information. All creatures that came before the Cyanozoic had to live without oxygen; they were anaerobes. From a chemical point of view this makes a lot of sense because oxygen is a very toxic poison: it reacts violently with most organic compounds, and destroys their properties when it is allowed free access to them. Like snakes with their own venom, the cells of most present creatures had to develop very sophisticated ways of handling oxygen before they could exploit its energy.

In the beginning, free oxygen would have been a deadly threat to life, and the evidence that it was not present on earth in the first two billion years confirms what chemists had suspected for a long time: that life originated and expanded in anaerobic conditions.

The first cells and all their descendants up to the oxygen revolution were therefore anaerobes. Let us call them collectively *archaeans*, and give the name of *Archaeozoic* to the period in which they lived, roughly from 3.6 to 2.6 billion years ago, between Prezoic and Cyanozoic.

Since bluebacteria were probably the first creatures which released free oxygen, some have suggested that they used it to exterminate competitors, but there are legitimate doubts about this. In the Archaean oceans there were plenty of ferrous ions to capture oxygen, and the removal of those ions took at least a few hundred millions years. The success of the bluebacteria — which is well documented — was more likely due to their immense versatility than to their power to exterminate competitors.

Later, when all ferrous ions were removed, oxygen could have been used as a weapon, but even then most of it escaped into the atmosphere and started accumulating there. The actual increase in the level of oxygen in the oceans was extremely slow, and gave many types of microorganisms plenty of time to adapt and develop the same defence mechanisms that the bluebacteria had acquired first. If oxygen was used in a chemical war, therefore, it was a war that many different types of cells managed to survive.

The transition from Archaeozoic to Cyanozoic was a very gradual affair and the date which separates them — 2.6 billion years ago — does not represent a discontinuity in the fossil record. It is rather a symbolic parting of two periods in which life as a whole faced different historical problems.

The Archaeozoic was the age that began with the origin of life itself, and during which the cells had to solve problems like surviving, diversifying and expanding in a totally anaerobic environment. The Cyanozoic instead was an age of plenty, an age of immense and immensely slow changes during which the microorganisms transformed the earth into an aerobic planet, and in the process transformed themselves.

Prezoic, Archaeozoic and Cyanozoic lasted approximately one billion years each, and represent the first three great chapters of the history of life (Figure 5).

THE PRIMORDIAL SOUP

Up to the 1890s, the growth of microorganisms in vitro was obtained by cultivating them in meat broths or in other organic solutions; the simplest medium devised by Pasteur, for example, contained ammonium salts, yeast ash and candy sugar. This gave origin to the idea that the oceans or the lakes where life appeared had to be full of nutrients, and formed what became known as 'the primitive broth' or 'the primordial soup'.

In 1899, however, Winogradski discovered that some bacteria can grow with only carbon dioxide and inorganic salts. Iron-bacteria grow on iron salts, sulphur-bacteria on sulphates or sulphides, nitrogen-bacteria on nitrites, and so on. All creatures which have the ability to synthesize their organic molecules from inorganic ones were given the name of 'autotrophs' (self-sufficient), while the others became known as 'heterotrophs' (food-dependent).

Winogradski's discovery provided an alternative to the primordial soup; it was now possible to think that the first cells were autotrophs, and that they appeared on a sterile earth. This theory, championed by Osborn (1918), Perrier (1920) and Constantin (1923), became fairly popular at the time because it avoids a substantial obstacle. If the first cells were heterotrophs, they had a deadline: once they had devoured all the nutrients in the primitive solutions, they would have starved to death. The theory of the ancestral autotrophs avoided this problem, but with the years it has become less and less convincing.

The cells obtain energy by three mechanisms — fermentation, photosynthesis and respiration — and fermentation is by far the simplest of them. More important, photosynthesis and respiration are long chains of reactions which begin with the short chain of fermentation.

This is not direct historical evidence, but it is evidence all the same and its message is clear: fermentation came first, while photosynthesis and respiration were developed later by adding new steps to the basic chain of fermentation. And since the fermenters had to find something to ferment around them, we conclude that the primitive solutions had to be full of organic molecules. The idea of the primordial soup becomes as necessary as the idea that plants must have minerals and animals must have plants around them.

There is also another experimental argument in favour of the primordial soup. Since the classic experiments of Stanley Miller in 1953, the formation of organic molecules in solutions has been studied extensively, and one major result has emerged: if the solutions contain free oxygen no biological compound is formed; if they are anaerobic, instead, it is relatively easy to obtain organic molecules. Most importantly, there are crucial compounds (like formaldehyde and hydrogen cyanide) which have been obtained in a very wide range of experimental conditions. In fact it has become virtually impossible to describe an anaerobic environment which simulates primitive conditions and where simple organic molecules cannot be formed spontaneously. From a chemical point of view, in other words, the existence of the primordial soup on the early earth is supported by fairly good arguments.

We have therefore three independent lines of evidence which converge to the same conclusion. The Iron Bands tell us that there was no free oxygen on the primitive earth; chemistry tells us that without free oxygen the ancestral solutions were bound to contain a variety of organic molecules; and biochemistry tells us that fermentation is the simplest mechanism that a cell can use to make a living, and the starting point of all the other metabolic mechanisms. As Pasteur said in 1875: "Fermentation is the consequence of life without free oxygen." We conclude therefore that the first cells had three characteristics: they were anaerobes, heterotrophs and fermenters.

THE INVENTION OF PHOTOSYNTHESIS

If the first cells were fermenters they were living, as we have seen, under the threat of extinction because their reservoir of food was constantly decreasing. Some of their descendants had to learn to build organic compounds from inorganic matter, but even that was not enough. The limiting factor in a growing population of microorganisms is free energy, and the only unlimited source of energy for them was sunlight.

The descendants of the first microorganisms had to invent photosynthesis, and this brings us up against a paradox: how could primitive cells have invented a process as complicated as photosynthesis, if fermentation allowed them only to *dismantle* organic molecules, not to build them? We can hardly say that they were thinking ahead, and we cannot assume that photosynthesis appeared suddenly because there are too many steps in it. There is however a logical solution, based on the fact that respiration is virtually the reverse of photosynthesis. This is not exactly true in all details, but on the whole the two mechanisms do use the same basic reactions in different orders. If respiration was developed first, therefore, photosynthesis would have required only a rearrangement of its reactions, and would no longer be unthinkable.

The important point is that fermentation and respiration are catalytic processes, they both dismantle molecules, while photosynthesis goes in the opposite direction. The fermenters had no use for light, and therefore no incentive to create photosynthesis, but they could definitely benefit from any improvement of fermentation, and respiration is just that, an extension and an improvement of fermentation.

The descendants of the first cells had a natural incentive to specialize, because the primitive solutions contained different types of organic molecules that could be used for food. Furthermore, the composition of the oceans was changing; the dead bodies of the cells were a net loss of organic matter, a new and increasing reservoir of food. There were therefore strong incentives to diversify, and the natural way of doing it was to add new catalytic reactions to the existing mechanism of fermentation, producing intermediate types of respiration.

Until a few years ago, this logical scenario was not considered realistic by the biochemists, because they believed that true respiration appeared after photosynthesis, not before. This is still valid, but true respiration is not the only respiratory mechanism that can exist. The clarification of this issue came in 1979, when Broda and Peschek showed that there is an intermediate logical step between fermentation and photosynthesis, and that this is precisely a form of respiration; they called it pre-respiration. From a biochemical point of view, the natural sequence was: fermentation → pre-respiration → photosynthesis → true respiration. The hypothesis that the first cells were fermenters, therefore, stands, and gives us our first glimpse of what happened in the beginning.

Clumsy little creatures, totally dependent for food on their environment and capable only of fermentation, appeared in primeval anaerobic solutions that were rich enough in organic molecules to feed them for many generations to come; their descendants learned to ferment a variety of different compounds, and some of them added new steps to the basic chain of fermentation, producing intermediate types of pre-respiration; finally, a few rearranged the pre-respiration reactions and obtained the first crude form of photosynthesis. When that happened, the microorganisms scored a major victory: they had managed to capture the energy of the sun.

THE INVENTION OF THE NATURAL CYCLES

We have seen that organisms can be regarded as either autotrophs or heterotrophs, but this classification is not entirely satisfactory because it does not mention the ability to decompose organic matter completely. A more accurate classification describes three types of creatures in Nature.

The *producers* (plants) make organic molecules from inorganic ones. The *reducers* or decomposers (fungi) dismantle organic matter and return it to the inorganic world. The *consumers* (animals) have an intermediate role; they feed on organic material, like fungi do, but their waste is still organic and has therefore to be treated by fungi in order to be recycled (in principle the animals are dispensable; life could well go on with only plants and fungi).

In the great cycle of production and decomposition, photo-synthesis is the first step in the production chain, and therefore is absolutely essential, but other steps of the cycle are no less vital. Since the resources of the earth are limited, the bodies of dead creatures must be decomposed and recycled, otherwise they would progressively deplete the world of some essential com-ponent, and life would come to an end even with an unlimited supply of sunlight.

The ancestral microorganisms therefore had to give origin not only to photosynthetic bacteria, but also to iron-bacteria, nitrogen-bacteria, sulphur-bacteria and so on. The cycle of life is a cycle of cycles: of carbon, of oxygen, of nitrogen, of phos-phorus, of sulphur, and so on. Every cycle consists of several reactions, and there is no living organism which can perform them all. Life, as we know it on earth, necessarily requires a com-munity of complementary creatures.

We realize in this way that the theory of Osborn, Perrier and Constantin which tried to exorcise the threat of extinction by assuming that the first cells were autotrophs, was based on a fal-lacy. Autotrophy alone does not guarantee survival. On the con-trary, if the first cells had been bluebacteria, for example, the threat of extinction would have been much greater.

The bluebacteria are immensely versatile, they adapt to virtu-ally all known environments and are the oldest survivors on earth. But precisely because they are so successful in conserving their kind, we cannot imagine them producing anything but other bluebacteria, and if they had been alone this resilience would have led them to extinction. The most important charac-teristic of the common ancestor was the ability to produce genuinely diverse descendants, and the bluebacteria are too sophisticated for that. Only a very imperfect creature was good enough to begin life on earth.

In the long run, at any rate, extinction could be avoided only by the self-perpetuating chains of the natural cycles, and the microorganisms therefore had a deadline not only for photo-synthesis but for all the other essential steps of the cycles.

They succeeded, of course, but we will attempt to understand how only later in the book. For the time being, let us say that the invention of photosynthesis was part of a more general process that we may call *the invention of the natural cycles*. If we make a list of the great problems that life had to solve, what we have seen so far gives us the first two entries: problem number one was the Origin of Life itself; problem number two was the Origin of the Natural Cycles.

PROKARYOTES AND EUKARYOTES

In order to appreciate the other problems that life had to solve in the Precambrian, we need a few concepts from the history of science, starting from observations made by the microscopists of the 18th century. By studying liquid suspension of micro-organisms, they noticed that all cells look like little pieces of jelly under the microscope. In some cases this jelly was perfectly smooth and transparent, while in others they could see a lump, or a dense body, inside it. They called this body *nucleus*, while the transparent jelly around it became known as *cytoplasm*. In 1774, Fontana added that the nucleus is not homogeneous, and described a corpuscle in its interior that became known as *nucleolus*.

The realization that there are two types of cells in Nature, some with and some without nuclei, emerged slowly from a great number of observations, and we cannot attribute it to any one scientist. Leeuwenhoek, for example, made drawings of both types without differentiating them, while a century later, when Fontana discovered the nucleolus, the existence of nucleated and non-nucleated cells was well known to microscopists, even if nobody seemed to attach great importance to it.

In 1839, Theodor Schwann (a biologist) and Matthias Schleiden (a botanist) published the papers which are said to mark the origin of the cell theory (the doctrine that all living creatures are made of cells), but in reality the theory had circulated long before that date. Thirty years earlier, for example, Lamarck had written "No body can have life if its parts are not cellular tissue, or are not formed by cellular tissue." What Schwann and Schleiden did contribute, instead, was a particular and very important form of the cell theory: they proposed that all animals and plants are made of cells which contain nuclei.

This idea has been tested countless times, and has become one of the most fundamental concepts of Biology. Its deep meaning is that only a cell with a nucleus can undergo embryonic development and give origin to visible creatures.

For some reason, cells without nuclei never break the microscopic barrier. They form that particular class of microorganisms that we call *bacteria*, most of which are single cells, but even their few multicellular forms (actinobacteria and myxobacteria, for example) invariably have microscopic dimensions and no embryonic development. Visible life belongs exclusively to nucleated cells, while in the microscopic world there are both types.

There are at present different classification systems in Biology, but in this book I will follow the convention that gives the taxonomic name of *Protista* to all nucleated microorganisms, and *Monera* to all non-nucleated or bacterial microorganisms. Other basic terms were introduced in 1957 by E.C. Dougherty, who divided the cells into two great classes: prokaryotes (without nuclei) and eukaryotes (well-nucleated). With this terminology, all bacteria are prokaryotic, while all other creatures (protista, plants, fungi and animals) are eukaryotic.

The concept that life exists on our planet in two distinct cellular forms is still debated, and later we will have to discuss it in detail, because it is crucial to our understanding of the history of life. For the time being, let us say that the first evolutionary theory which is specifically based on this concept was proposed in 1866 by Ernst Haeckel, the man who classified all bacteria into one group and gave them the taxonomic name of Monera.

The theory of Haeckel is straightforward. The cells which lack nuclei are the simplest living creatures, and therefore the most primitive ones: the first cells were bacteria. According to Haeckel, life evolved in three stages: first the bacteria appeared, then some bacteria gave origin to protista, and finally some protista gave origin to animals and plants. Today this is referred to as *the Prokaryotic theory* of cellular evolution, and the three steps of Haeckel are written in the sequence: prokaryotes → monocellular eukaryotes → multicellular eukaryotes.

Using these concepts, let us now return to the Precambrian and to the reconstruction of its main events.

THE AGE OF THE PROTISTA

From the end of the Cyanozoic to the Cambrian explosion of life — roughly from 1.6 to 0.6 billion years ago — we find a novelty in the fossil record: the average dimension of the cells is greater than in all previous periods.

From calculations performed on spheroidal forms, the Archaeozoic and Cyanozoic microfossils have the diameters of small and medium-size cells. Most of them (over 50%) are less than 5 microns, and none is greater than 100 microns (a micron is a thousandth of a millimeter). In the last billion years of the Precambrian, instead, the most frequent forms are between 5 and 20 microns, and there is a significant percentage of cells which exceed 100 microns; a few of them are even greater than a millimeter (Schopf, 1978).

We know that prokaryotic cells, on average, are definitely smaller than eukaryotic ones; a typical prokaryote is even smaller than the nucleus of a typical eukaryote. The actual sizes of both types, however, vary considerably and there is a range — between 5 and 20 microns — where their dimensions overlap. A very substantial percentage of Archaeozoic and Cyanozoic microfossils have dimensions in this range, and could be therefore either type, but there is a biochemical argument that tells us otherwise.

Virtually all eukaryotes need oxygen to live, and even the very few exceptions to this rule seem to be the descendants of oxygen-dependent creatures which adapted later to anaerobic life. Among the prokaryotes, instead, we find a wide variety of responses to oxygen. Some bacteria cannot tolerate it, they are obligate anaerobes; others can survive without it (facultative anaerobes), and a third class depend upon it absolutely, like most eukaryotes do (obligate aerobes).

These characteristics suggest that some prokaryotes evolved when there was no oxygen on earth and others when the concentration of oxygen was changing, while true eukaryotes appeared only when the level of oxgyen became stable and relatively high.

Using this argument, the increase in cell size that we find in the fossil record precisely from the end of the oxgyen revolution onwards, has an unambiguous meaning: it signals the appearance of the first true eukaryotic microorganisms, the origin of the protista. From what we see in the fossil record, the last billion years of the Precambrian was characterized by the appearance and by the planetary expansion of these microorganisms, and will be called therefore *the Age of the protista* or, by an old name, *Proterozoic.*

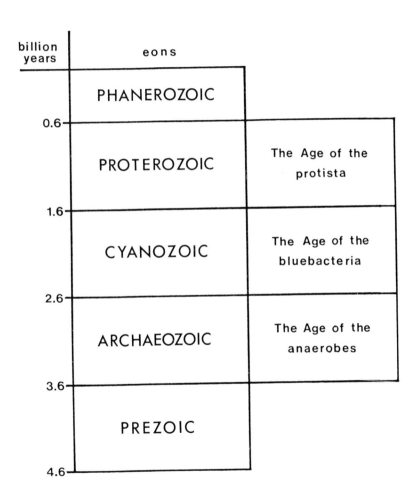

FIGURE 5:   The five eons of the history of life.

The other great event in this period was the appearance, towards the end of it, of the first multicellular creatures. We cannot put a precise date on this event, but the record shows that there was a very substantial gap (probably more than 500 million years, and certainly not much less) between the first protista and the first multicellular organisms. The origin of the eukaryotic cell and the origin of multicellularity were definitely two separate historical events.

We can divide therefore the history of the earth, and of life on earth, into 5 great ages or eons: four ages of one billion years each in the Precambrian, and a last chapter of 0.6 billion years afterwards (Figure 5).

THE PROBLEMS OF EVOLUTION

Every one of the four precambrian ages — Prezoic, Archaeozoic, Cyanozoic and Proterozoic — opened a new chapter in the history of life. The first led to the origin of the cell, the second to the oxygen revolution, the third to the origin of the eukaryotes, and the fourth to multicellularity and visible life.

At the end of the previous chapter we concluded that the fundamental developments of evolution took place in the Precambrian, but now we can be a little more specific. In a way, it is Nature itself that tells us what a modern theory of evolution should explain.

The origin of the cell, the origin of the natural cycles, the origin of the eukaryotes, and the origin of multicellularity were the four main steps in the history of life, and they become therefore, almost by definition, the four problems of evolution (the fifth great event was the origin of the mind, at the end of the Phanerozoic, but this book will not deal with it). Now that we know where we want to go, let us discuss how to get there.

We have already seen that important conclusions can be obtained by combining the information of the fossil records with biochemistry, but this procedure is not entirely fool-proof. The conclusion that the first cells were fermenters, for example, was obtained from biochemical arguments, but in reality it is valid only if we assume a slow evolution on earth for them. Historically they could have appeared suddenly, for example they could have come from space, and the above conclusion would be totally wrong. The same could be true for other cells, for example for the first eukaryotes, and in this case there would be no point in trying to understand how these creatures evolved from preexisting ones.

In experiments which simulate the conditions on the primitive earth, a variety of organic molecules have been formed, but in the last thirty years radioastronomers have found that most of these molecules also exist in space, together with many other organic compounds. In these circumstances, it would be unwise to restrict our search to a terrestrial theory.

Cosmic life may exist now, could have existed in the past, and could have contributed to terrestrial life. Before discussing anything else therefore, we must try and find out if this contribution is likely to have been substantial or negligible. More precisely, we would like to know if and how Cosmic Life had a role in the four great precambrian events that shaped the history of life on earth.

CHAPTER 3

Theories of Sudden Life

THE IDEA OF ETERNAL LIFE

In 1862 Lord Kelvin estimated the earth to be a mere 100 million years old, and later he actually reduced the estimate to about 20 million. This meant that there had not been sufficient time on earth for a slow evolution of the first cells from inorganic matter.

In 1860, on the other hand, Pasteur proved that the sudden generation of microorganisms in an organic solution is absurd. In these circumstances, the first cells could not have had a terrestrial origin, either slow or quick, and there were only two alternatives to choose from: either they had been created by God, or they had arrived on earth from Space. Darwin opted for the first possibility, Lord Kelvin and many others for the second.

The idea of life from space, however, was not only a consequence of Lord Kelvin's calculations, but also the expression of a general view that was emerging at the time: the view that organic life is eternal, that there had been no beginning.

One must remember that the cornerstone of 19th century physics was the principle that matter and energy are neither created nor destroyed, and that was translated into the idea that matter is eternal. In this case life, as a property of matter, could be eternal as well.

Liebig (1861):

It is sufficient to admit that life is as old and as eternal as matter itself, and the entire argument about the origin of life loses apparently all sense.

Von Helmholtz (1874):

It is entirely within the domain of scientific discussion to enquire whether life had ever been created, whether it is not just as old as matter itself.

In this way, the problem of the Origins was divorced *in principle* from the problem of evolution.

"The beginning of life on earth is absolutely and infinitely beyond the range of sound speculation" wrote Lord Kelvin, and in a private letter Darwin expressed a similar opinion:

It is mere rubbish thinking at present of the origin of life; one might as well think of the origin of matter.

The idea of Life from the Universe grew in this climate: if life is eternal and our planet is too young for a long process of evolution, the first creatures must have come to earth from space. Eventually, the idea took a precise form: microorganisms travel through space and colonize every suitable planet on which they happen to fall. This theory became known as *Panspermia* (Cosmic Insemination) after the name proposed in 1908 by Svante Arrhenius, who was its most vigorous proponent.

As we will see, new versions of Panspermia have been put forward in recent years, but we should keep in mind that the very idea that gave origin to Panspermia — the idea of eternal organic life — has completely collapsed. The realization that the Universe is expanding, and that the expansion began with an immense explosion (the Big Bang), have buried the idea for good.

The Big Bang left a background radiation which can still be detected (some call it the whisper of Creation), but the tic-tac of a Geiger counter tells us, much more directly, the same story: the atoms that make up our bodies have not existed for ever. They were created within giant stars, and were scattered in space when these became Supernovae and exploded.

The age of the Universe is still debated. The estimates range from 8 to 20 billion years, with a distinct preference for the low-to-middle values; a figure between 12 and 16 billion years appears to be the most likely candidate. From the Big Bang to the explosion of the first Supernovae and the formation of the first planets, we must allow at least 1 or 2 billion years, and we conclude therefore that life could not have appeared, on any planet of the Universe, much earlier than 10 billion years ago.

The age of the Universe tells us two major things about life. One, as we have seen, is that there is no going back to the idea of eternal organic life. To deny the historical reality of the Origin of Life is tantamount to saying that radioactivity does not exist, that the Universe is not expanding, that physics has got virtually everything wrong. The second message is that the origin of the first cells must have taken place in less than 10 billion years, probably much less. The maximum *order of magnitude* that we can allow for it is therefore one billion years, wherever it happened.

We conclude that the origin of life was an historical event: it happened, somewhere in the Universe, between 3,5 and 10 billion years ago, and was either an instantaneous event or an evolutionary process that lasted one billion years or less, perhaps much less but certainly not much more.

COSMIC PANSPERMIA

In a recent series of books (Lifeclouds, 1978; Diseases from Space, 1979; Evolution from Space, 1981), Fred Hoyle and Chandra Wickramasinghe have reproposed the theory of Panspermia, but have introduced in it an element of such novelty that it is essential, in my opinion, to differentiate clearly between the old and the new version of the theory.

Arrhenius and his predecessors believed that life develops only on planets and that its seeds are somehow catapulted into space and dispersed throughout the Universe. Their journeys always begin on planets, and I will therefore refer to this view as *Planetary Panspermia.*

Hoyle and Wickramasinghe instead say that genes, viruses and bacteria (and perhaps insects) originate, grow and multiply in space, in the immense regions where apparently there is only cosmic dust. Here, space is not only the medium through which they journey, but the very womb which gave and still give birth to them, and I will therefore refer to this view as *Cosmic Panspermia.*

The first spectra of interstellar organic materials were discovered in 1937, and the list of organic molecules in space has steadily grown ever since. Following this line of research, Hoyle and Wickramasinghe found, in 1977, that the infrared spectrum of dry cellulose fits fairly well with the spectrum of the astronomical object known as OH26.5+0.6. Similar correspondences were found for other astronomical sources of infrared radiation, and led them to conclude that the interstellar grains are made of complex carbon compounds. When that was added to the finding that the grains have dimensions in the range of one micron, Hoyle and Wickramasinghe concluded that they are made of bacteria and viruses.

Most astrophysicists do not agree with any of their claims, and maintain that the interstellar grains are mixtures of inorganic substances, mainly silicates, but this point has not yet been settled conclusively and we must keep an open mind about it.

Hoyle and Wickramasinghe reject the Darwinian theory because this predicts a gradual transformation of forms, while the fossil record shows sudden transitions and discontinuities. This is a long-standing objection that deserves every consideration, but it is by no means original. What is truly new is the solution that Hoyle and Wickramasinghe have proposed.

The discontinuities of the record, they say, can be explained by the injection of fresh genes from space. The genetic pool of all organisms living at any one time is too limited and changes too slowly to account for the rapid increase in diversity and complexity that took place. But if genes are sent to earth from the enormously larger laboratory of space, we can easily account for their extraordinary diversity, and can explain naturally why evolution took place in jumps and jerks. The collective pool of the terrestrial genes has been periodically supplemented by cosmic genes in a prolonged and gigantic operation of genetic engineering from space.

The idea is certainly staggering. More important, Cosmic Panspermia can be tested and is therefore a genuine scientific hypothesis. It can be tested because if bacteria, viruses and genes account for the spectra of astronomical regions, their numbers and their swarming through space must be truly immense. This means that tons of bacteria keep pouring every year onto the solar system. Hoyle and Wickramasinghe calculate that no less than 10^{22} bacteria, or 10^{24} viruses, fall on the earth alone each year, and that equivalent amounts hit the moon and the other planets.

Unfortunately the moon has no atmosphere and the bacteria would be destroyed on impact, but perhaps a system able to detect the bacterial rain could be installed there. Better still, we could explore the surface of Mars, because there the bacteria would land softly, as they would do on earth and on any planet which has an atmosphere. If Cosmic Panspermia is true, the surface of Mars should be literally littered with genes, viruses and bacteria.

We know of course that Viking-1 and Viking-2 have found nothing of the kind, but Hoyle and Wickramasinghe are not bothered by this. They suggest that the spacecrafts simply landed on the wrong spots, and that we must wait for the results of future explorations. This is a little surprising, but future explorations of Mars will undoubtedly come and we might as well wait for them before writing Cosmic Panspermia off. .

Some say that Cosmic as well as Planetary Panspermia can be written off already, because no form of life can survive a direct prolonged exposure to cosmic radiation. Hoyle and Wickramasinghe have not accepted the argument, and it is fair to say that we need more information to settle this issue conclusively.

It is also fair to say that, while keeping an open mind for the unexpected, we cannot brush aside the evidence that we do have. The first two explorations of Mars, and what we know about the biological effects of the cosmic rays, are strong evidence against Planetary and Cosmic Panspermia, and I am not convinced by the argument that these are the only theories worthy of the Space Age, as Hoyle and Wickramasinghe have claimed.

DIRECTED PANSPERMIA

In 'Life Itself, its Origin and Nature' (1981), Francis Crick has championed a new version of Panspermia: life on earth was started by bacteria which arrived here nearly 4 billion years ago on board a spaceship sent by an advanced civilization.

Crick has named this hypothesis *Directed Panspermia*, and has frankly admitted that, as yet, we have no evidence for it. He maintains instead that the hypothesis is "not implausible" and "the kindest thing to state about Directed Panspermia is to conclude that it is indeed a valid scientific theory, but that as a theory it is premature". Let us examine it.

Since the Universe is about twice the age of the earth, "there is enough time for life to have evolved not just once, but two times in succession", all the way from microorganisms to a civilization capable of sending rockets into space. It is not implausible therefore, says Crick, that such a civilization evolved on another planet in the first half of the Universe's history, and in this case it could well have decided to launch rockets packed with bacteria and programmed to land on any suitable planet.

The rockets might have been sent as vanguards with the task of making planets hospitable before a new wave of advanced space colonization could follow. Alternatively, they could have been launched in the wake of an impending disaster. Suppose that those people discovered that they were doomed, for example that a neighbouring star was set on a collision course with their solar system. In that case they would have tried to escape, but a rocket designed to carry them, their descendants and their supplies for thousands or millions of years would have presented formidable problems.

As a precautionary measure, they could have decided to launch unmanned rockets containing only microorganisms to ensure that at least the most elementary forms of life would survive. These rockets would be simpler and smaller, the bacteria could be frozen and packed by the billions into tiny spaces, and they could be protected from cosmic radiation within suitable containers. In short, the project would have been desirable because "bacteria can always go further".

Crick describes the relationship between Directed Panspermia and the traditional terrestrial theory in this way:

For all that we know the location was vital ... it is rash to assume that conditions here were just as good as anywhere else. Whether life originated here or elsewhere is, at bottom, an historical fact, and we are not entitled, at this stage, to brush it aside as irrelevant. The two theories, then, are radically different.

Let us take a closer look at this difference. In practice, we could not build a theory valid for the earth alone even if we wanted to, because we do not have enough information about the primitive earth to do this. The objective of the traditional theory, therefore, is to study the origin of life in conditions which can be loosely defined as *earth-like*, and which apply a priori to all earth-like planets.

Life requires carbon compounds dissolved in water, and vast quantities of water can be prevented from flying off into space only if they are held by a suitable gravitational field, i.e. if they exist on a material body which has a convenient mass. Furthermore, life and liquid water require a restricted range of temperatures for long periods of time, and in our Universe this is obtained only if the body which carries them goes around a star at a distance which is neither too small nor too big in relation to its mass. This is the essential description of an earth-like planet, and Crick agrees that life must have originated on such a location.

Let us assume that life began on the planet suggested by Crick, and let us imagine ourselves interviewing the advanced people who lived there about *their* theory on the origin of life. What is certain is that they would not even consider Directed Panspermia as a suitable theory, because they had no other younger planet to go back to. They would either believe in Creation by God, or would study the origin of life by evolution on their earth-like planet as we are doing on ours. Our theory could be considerably different from theirs if we were to base it on *earth-specific* conditions, but since we do not know these conditions, such a danger is non-existent. Our earth-like theory (if we manage to build a consistent one) would be as similar to theirs on the origin of life as it would be on the origin of the Universe, on the origin of the stars and on the origin of the atoms.

Directed Panspermia is no substitute for an earth-like theory of the origins because it is not a theory of the origins. It is a theory on the transport of life from one earth-like planet to another, and has nothing to do with the process by which, somewhere in the Universe, life began.

LIFE IN THE UNIVERSE

Toward the end of the Renaissance, in 1584, Giordano Bruno spoke of "infinite worlds in the Universe inhabited by intelligent beings". The idea was seen as an arrogant defiance of authority and for this, and other heresies, Bruno was sent to the stake by the Inquisition in 1600.

The idea of infinite worlds, to be precise, was not new; the novelty was that Bruno claimed a scientific basis for it, and in his times this was dynamite. Colombus had discovered the new world, Leonardo had created the universal man, Copernicus had described the new solar system, and Galileo was about to found the new science. By saying that the new Universe is full of life,

Bruno was contributing to the revolution of the Renaissance with the most general principle of modern science: the principle that the laws of Nature are identical everywhere in the Universe, in space and time. There are infinite suns and infinite planets in the sky, he was saying, and since the laws of Nature are the same in all places, there must be infinite worlds populated by intelligent beings.

Today, we put it in this way. There are 100,000 million stars in our galaxy, and at least 10,000 million galaxies in the Universe. Many of the stars are likely to have planets, and a fraction of the planets are likely to be like the earth. Even if there is only one sun in a million stars, and one earth-like body in a million planets (a very conservative estimate), this still amounts to a billion earth-like planets in the Universe.

Furthermore, we do accept Bruno's principle that the laws of Nature are identical everywhere, and therefore arrive at a similar conclusion: life may well exist somewhere else in the Universe.

This idea is so modern that we regard it as a product of the Space Age, and Bruno is hardly mentioned. What is more important, however, is to notice that there is a fundamental difference between the theory of Bruno and the other theories of Cosmic Life. While all three versions of Panspermia assume that life came to earth from space, Bruno proposed that life arose independently on every inhabited planet. Bruno's idea was 'Life *in* the Universe', not 'Life *from* the Universe'.

The distinction is important, because the proponents of Panspermia have often used the psychological argument that evolution, as an earth-bound theory, is the last resort of the anthropomorphic view that our planet is special and we are unique. In reality, evolution is perfectly compatible with the vision of Bruno and, strictly speaking, should not be referred to as a *terrestrial* but as a *planetary* theory. Bruno gave a cosmic dimension to life before the theory of evolution was even conceived, and such a dimension has been there all the time, with or without Panspermia.

THE CREATION HYPOTHESIS

The idea of Creation has been expressed throughout history in innumerable forms, the latest of which is the hypothesis that God created the first microorganisms. This version may not look particularly inspiring, but it is undoubtedly original, and has the advantage of being compatible with the fossil record.

The general public seems unaware that this Creation hypothesis was first formulated by Darwin, probably because it is one of the few Darwinian ideas that scientists have not taken seriously. They have seen in it an illogical formality, a mere concession to the Church or to religious feelings in general.

And yet Darwin stood by the idea throughout his life:

I imagine that probably all organic beings which ever lived on this earth descended from some primitive form which was first called into life by the Creator.

Furthermore, he repeatedly dismissed the attempts to extend evolution to the origin of life:

Evolution has nothing to do with the origin of the soul, nor with that of life itself.

Was this only a public-relations act? a twist of Victorian morality?

Let us examine an alternative. Suppose that Darwin did not believe that matter is eternal, that he was really convinced that *evolution needs a physical beginning*, because nothing would still be evolving if the Universe had existed for ever. To suggest that matter may not be eternal was equivalent, in his times, to scientific suicide, and Darwin had no taste for that. The hypothesis that God created the first cells, on the other hand, was an ideal solution: nobody would take it seriously, but it would still amount to saying that Lord Kelvin and comrades were wrong, that evolution is incompatible with eternal organic life and with eternal matter.

But let us return to the hypothesis itself. Whatever reason Darwin had in mind, it is an historical fact that he proposed it first, and we should discuss it for what it is, an original hypothesis on the origin of the first cells. More precisely, we are interested in examining its experimental implications. What characteristics would we find in the fossil record if the first microorganisms were created by God?

Francis Crick proposed the following:

The main difference would be that microorganisms should appear here suddenly, without any evidence for prebiotic systems or very primitive organisms. We might also expect that not one but several types of microorganisms would appear.

Crick was speaking of the implications of Directed Panspermia, but that doesn't matter. The important point is that the Creation hypothesis of Darwin, the Spontaneous Generation of Lamarck, Planetary Panspermia, Cosmic Panspermia and Directed Panspermia predict exactly the same experimental pattern. They are five different versions of one idea: the idea of Sudden Life.

In these circumstances, our priority is no longer to discuss which version of Sudden Life is most plausible, but to examine their common denominator, to find out if life did or did not appear suddenly on earth.

THE EARLY EARTH

The idea of Sudden Life could be tested by collecting a significant number of samples from the primitive crust, and checking if the rocks which preceeded the oldest fossilized cells have any trace of prebiological activity. Unfortunately, we cannot perform this test, because whatever crust existed on our planet in the first 800 million years seems to have been swallowed into the interior of the earth and melted down completely.

There is however a second possibility. If we could prove that the surface of the early earth was an inhospitable inferno, and that microorganisms appeared soon after the environment became tolerable, we would have a very strong case for the sudden origin of life. This is the actual reasoning which is used today by some scientists to conclude either that the first cells appeared instantaneously, or that they evolved very quickly, in a few million years or less.

In 1979, for example, Cyril Ponnamperuma found in the Isua rocks traces which could be interpreted as the result of photosynthesis, and in 1980 he announced the following conclusion:

We have now what we believe is strong evidence for life on earth 3,800 million years ago. This brings the theory for the Origin of Life on Earth down to a very narrow range. Allowing half a billion years (for the disturbed conditions described above) we are now thinking, in geochemical terms, of instant life.

The signs of a photosynthesis-like activity could have been left in the Isua rocks by microorganisms as well as by precellular systems, but this is hardly the main issue. The crucial point is the assumption that in the very long period which preceeded the Isua rocks, the earth had predominantly been a hostile place, a sterile and sterilizing planet where any hint of life would have been quickly destroyed.

This is a scenario for which we are well prepared, because virtually all textbooks on the subject represent the primitive earth as a red-hot cauldron of boiling lava, the nearest thing to hell that one could think of. As a result, various authors have given the first 600 or 700 million years the name of *Hadean*, from Hades, the Greek God who ruled over the dead. Perhaps it is a little illogical to associate with death a period which preceeded life, but the name is appropriate enough if one believes that the primitive earth was a physical inferno. This, however, is precisely the point where opinions are divided.

In a paper which appeared in Nature in 1982, Henderson-Sellers and Graham Cogley summarized the results obtained by various scientists and by their own team with a scenario which is very different from the traditional one. The new picture is that of a relatively cold planet whose average temperature was just above freezing, and which had been almost totally covered by water since the beginning, that is to say since the formation of an early crust, nearly 4.6 billion years ago.

The paper ended with the following conclusion:

A substantial early hydrosphere, then, is both consistent with results from a variety of disciplines and, as our results show, powerful in solving the paradox of the faint young Sun. The most plausible date for its origin is at the very beginning. Most of its properties, and especially its stability against large, irreversible changes of composition and climate, seem to follow if we identify the principal surface volatile as liquid water. There was thus a long expanse of pre-Isuan time during which the Earth could have sheltered and nurtured organisms or their precursors.

This new scenario does not imply that the primitive earth was an undisturbed place. Volcanic eruptions, earthquakes, cataclysms, intense radioactivity, meteoric bombardment and the like are all regarded as very realistic occurrences, but the important point is that they were not incompatible with the existence of vast expanses of liquid water.

We have therefore in front of us two very different pictures of the early earth: one represents it as a red-hot hell, while the other comes nearer to the blue image of a planet bathed in water virtually since the beginning. The conclusion that we must think in terms of *instant life* therefore is by no means inevitable.

The age of the Universe, as we have seen, implies that the maximum order of magnitude that we can allow for precellular evolution is one billion years anywhere in the Universe, and on the early earth we had just that. From 4.6 to 3.6 billion years ago there was the time, and perhaps there were the conditions. This means only that life *could* have evolved on earth, not that it did, but in our present state of knowledge that is all we can ask for.

AN EPILOGUE AND A PROLOGUE

Lamarck and the evolutionists of the 18th century believed in the sudden generation of microorganisms; Darwin proposed that the first forms were suddenly created by God; Liebig and Arrhenius supported Panspermia, and so did Hoyle and Wickramasinghe and Crick. And all the philosophies and religions of the past delivered the same message: the first creatures appeared suddenly on earth, created either by God or by Mother Earth.

Sudden Life, in one form or another, has been the only solution to the problem of the Origins that the human mind had conceived throughout history, up to the end of the 19th century. The alternative idea that the first creatures evolved from lower forms does not even have a birthdate and a paternity. It circulated anonymously, in the form of rumours, in the second half of the 19th century, and we only know of its existence because of the public attempts to dismiss it. Lord Kelvin was an outspoken opponent of the idea, and Darwin made it clear that he did not like it.

Today, the idea of prebiotic evolution is accepted, at least in principle, by many scientists, but we should remember that we still have no direct historical evidence for it. In the Isua rocks and

in other very old sediments there are traces that could mean prebiological activity, but they are not certain. Even the geological
theories of the early earth, as we have seen, do not give us a clear-
cut answer: we can only say that prebiotic evolution could have
taken place.

In these circumstances, it is premature to dismiss the idea of
Sudden Life in any of its terrestrial or cosmic versions. If this is
true, however, it is also true that we do not have to prove Sudden
Life wrong in order to consider precellular evolution. On the
contrary, we must examine this alternative as thoroughly as
possible, if we want to keep an open mind on Nature and on the
history of Nature.

From what we know at present, there is no evidence that
Cosmic Life made a substantial contribution to terrestrial life,
and in particular to the four great precambrian events that
shaped the history of life on earth. In the rest of this book, therefore, I will assume that these events were, first and foremost,
terrestrial processes.

CHAPTER 4

Genotype and Phenotype

MENDEL AND THE CLASSICAL THEORY OF EVOLUTION

Since our reconstruction of the history of the cell depends heavily on our understanding of the cell itself, I will dedicate this chapter to a brief review of cell biology, and will return to the problem of the Origins only in the next chapter. Let us start by discussing heredity, the property which is the very symbol of life.

In heredity, Darwin's theory of Pangenesis was little more than a restatement of an old Hippocratic idea: in the reproductive system there are particles from all parts of the body which make a child's organ grow as a blend of his parents' organs. The theory explains why children resemble their parents but, as Fleeming Jenkin pointed out, has a serious drawback in evolution. If hereditary characters are blended, any novelty or variation in them would practically disappear in a few generations.

A yellow pea in a population of green peas, for example, would give origin to yellow-green descendants in the first generation, and then to descendants which step by step go back to the original green colour. Whatever the advantages of the yellow variation, in time the character would just fade away, like a drop of ink in the sea.

Darwin's theory of evolution, however, required the opposite: if a new character is useful, it should be transmitted in full to more and more descendants, and eventually it should replace the old one. But Jenkin's objection was valid, and Darwin became aware that his theory of evolution badly needed a consistent theory of heredity if it was to survive.

The irony is that Darwin had the solution in his own house, in a paper which had been sent to him by an unknown author, but he was unaware of it; after his death, the pages of that paper were to be found uncut (and they still are). In that way he prevented himself from reading the very words which were going to save his beloved theory of evolution.

The unknown author was Gregor Mendel, who had sent his paper to various other leading scientists only to see it either ignored or dismissed with contempt, and who died in total obscurity in 1884, two years after Darwin.

In one of his experiments, Mendel crossed yellow and green peas for many generations but never obtained an intermediate colour. Sometimes a green pea was produced by parents which were both yellow, but the important point was that mixing or blending was never observed. Equally important was the fact that the proportions of yellows and greens followed simple mathematical combinations.

Mendel built a model. He assumed that for every hereditary character there are two factors (one from each parent), and that the actual character is the full expression of only one of them. If they are different, the factor which is expressed is called *dominant* and the other *recessive*. For convenience, let us represent a character with two letters, a capital letter for the dominant factor and a small letter for the recessive one, "A" for yellow and "a" for green, for example.

If the characters of the parents are AA and aa, the offspring will get an A from one parent and an a from the other, and will all be Aa, all yellow. If the parents are Aa and Aa, they will both produce factors A and a, and the offspring would have any of the four possible combinations, AA, Aa, aA, aa; statistically, three quarters would be yellow and one quarter green. The crossing of Aa with aa would allow the combinations Aa, aA, aa, aa: half yellow and half green, while the crossing of aa with aa would obviously produce only green, and the crossing of AA with AA only yellow.

In his experiments, Mendel not only obtained all the combinations predicted by the model, but found that these were the only ones which could in fact be produced. And he obtained the same results with other characters as well, for example with the form of the seeds (smooth or wrinkled), and with the colour of the flowers (white or red). Furthermore, the model accounted for all the results obtained in successive generations, and there was no doubt therefore that, as far as peas were concerned, the evidence was strong.

The implications of Mendel's model are straightforward and revolutionary. Heredity is carried in units which maintain their individuality and which are either fully expressed or fully ignored: it is an all-or-nothing phenomenon, with no intermediates, no mixing and no blending. Despite this, the model can easily explain why children resemble both their parents; one has only to assume that most visible characteristics are determined not by a single factor but by many of them. If some dominant factors come from the father and some from the mother, the final result would just look like a blending of the parental characteristics. In evolution, the Mendelian model implies that if a novelty appears, the new factor can be transmitted intact to future generations, and can replace the old one. This is precisely what Darwin required but had been unable to account for.

The general public does not seem aware that the modern theory of evolution by natural selection is not Darwin's original model, but a synthesis of Darwinism and Mendelism, a theory created in the 1930s by Ronald Fisher (1930), Sewall Wright (1931), John B.S. Haldane (1932) and Theodosius Dobzhansky (1937), and extended in the 1940s by Julian Huxley (1942), Ernst Mayr (1942) and George Gaylord Simpson (1944). This unified theory has been given various names such as *The Modern Synthesis*, *The Synthetic Theory* or *The Genetical Theory of Natural Selection*, but since it is reminiscent of the 19th century synthesis of Mechanics and Thermodynamics which gave origin to Classical Physics, I will call it *The Classical Theory of Evolution*.

GENOTYPE AND PHENOTYPE

Mendel's ideas were rediscovered in 1900 by Hugo de Vries, Carl Correns and Erich von Tschermak, but for at least ten years most biologists opposed them with the same fervour with which later they were going to support and endorse them. The collective conversion to Mendelism of the biological community has been described by the historian Garland Allen in the following way:

What many workers between 1900 and 1910 failed to grasp was the fundamental distinction between the hereditary particle itself and the recognizable adult character to which it presumably gave rise. This distinction between what is called today the *genotype* and the *phenotype*, was explicitly pointed out in 1911 by the Danish botanist Wilhelm Johannsen. Johannsen emphasized that organisms did not inherit 'characters' at fertilization, but only specific genetic components, or *potentialities* for those characters.

The *inherited potentiality* he called the *genotype*. It represented (1) what the organism could pass on to the next generation, and (2) what the organism itself *could* (but not necessarily *did*) show as a visible adult character.

The distinction was crucial, since failure to make it led many workers — especially embriologists such as Morgan — into a quandary. Unconsciously, they saw phrases such as 'inheritance of dark coat colour' as implying that a sperm or egg carried the dark coat colour condition. And such talk sounded too much like preformationism. Johannsen's distinction made it possible to think of hereditary particles not as fully developed adult characters, but as units guiding functional processes. If a 'gene' (Johannsen introduced the term in 1909) was only the potential producer of an adult character, then the question of how that adult character was actually formed became a question open to study. Furthermore, Johannsen's conception made it possible to understand more clearly the Mendelian idea of dominance and recessiveness. Once the distinction became clear in the minds of those workers concerned with heredity and evolution, it became easier to accept the Mendelian theory in a modern context.

The event which prepared the ground for the acceptance of Mendelism, and later for the synthesis of Darwinism and Mendelism, was therefore the introduction in Biology of genotype and phenotype. These concepts have been with us ever since, unscathed by time.

Ernst Mayr (1978):

Every biological individual has a peculiarly dualistic nature. It consists of a genotype (its full complement of genes, not all of which may be expressed), and a phenotype (the organism that results from the translation of genes in the genotype).

The idea that the cell and every living creature is a duality of genotype and phenotype has become the central concept of modern biology, the most synthetic way of describing anything that lives, the very definition of life. It is like saying that a human being is a duality of mind and body, or that a computer is a duality of software and hardware.

The idea is so general that there is perhaps the risk of regarding it as an abstraction too much detached from the reality of life, but a quick look at the actual components of the duality will show that genotype and phenotype are concepts full of biological meaning, that they truly represent the synthesis of an enormous number of experiments and meditations.

PROTEINS

In general, solids are turned into liquids by heating them, and the process is reversed by cooling, but biological substances do not behave like that. Since the 18th century, chemists noticed that egg-white, milk and blood, among other things, are coagulated by heat. They turn from liquid to solid, not the other way round, and the process is irreversible. The components of life seemed to have a chemistry of their own.

In 1777, Pierre Macquer gave the name of *albuminoids* to all substances that are coagulated by heat, and regarded them as the chemicals of life, a concept that gained wide popularity. In 1838, Gerardus Mulder extracted from various albuminoids a substance that he regarded as their common factor, and obtained for it the empirical formula $C_{40}H_{62}O_{12}N_{10}$. He announced that this substance

is without doubt the most important of the known components of living matter,

and to make sure that the message would not be lost he named it *protein*, from the Greek 'proteios' which means 'of the highest importance'.

In reality, Mulder had obtained a mixture of different things, not a pure substance, but the name remained. While most organic compounds like sugars, starch and fats contained only carbon, hydrogen and oxygen, substances like those obtained by Mulder were also rich in nitrogen or nitrogen and sulphur, were very complex, and could be extracted from all living organisms. For these reasons, chemists put them in a class of their own, a class of life-specific chemicals, and called them proteins.

Later, they found a better definition. It was discovered that continuous boiling in acid or alkaline solutions splits proteins into units, or building blocks, that turned out to be aminoacids, molecules which have an amino group (NH_2) at one end, and an acid group (COOH) at the other. (There are at least 20 types of aminoacids in Nature, and what differentiates them is a third chemical group, called radical, which represent the identity label of each type.) Proteins therefore could be defined as 'giant clusters of molecules, or macromolecules, which, upon hydrolysis, yield aminoacids'.

At the beginning of our century, Emil Fischer showed that any two aminoacids can be linked together by eliminating a molecule of water from the bridge formed by the acid group of one aminoacid and the amino group of the other. In this way, the bridge COOH-HNH becomes CO-NH, linking the aminoacids in a strong chemical bond that Fischer called a 'peptide bond'. Two aminoacids linked in this way are known as a peptide, and longer chains as a polypeptide.

Fisher managed to build small polypeptides in the laboratory, up to 18 aminoacids long, and proposed that all proteins are built on the same principle: they are linear chains of aminoacids joined by peptide bonds. He could not prove this theory because none of his small polypeptides behaved like proteins, but built a strong case for it.

At the same time, Fischer proposed that proteins have a precise three-dimensional structure, i.e. that each type has a number and an arrangement of atoms in space which is stable and specific for that type. In short he proposed that proteins are molecules like all the others, bigger in size and more complex in behaviour, but molecules all the same, with a precise chemical composition and, as a consequence, precise physico-chemical characteristics.

He had a reason for this. Crystals of hemoglobin had been discovered in 1840 (by Hünefeld), had been grown in vitro in 1850 (by Funke), and in 1870 they had been obtained from at least 50 different species, man included. In a crystal all molecules are identical, or very nearly so, and the mere fact that proteins can crystallize means that they have a precise arrangement of atoms in space.

Fischer's theory that proteins consist of linear chains of aminoacids and have specific three-dimensional structures, was therefore plausible, and most biochemists accepted that there could be some truth in it. But was it the whole truth? It was all very well to say that 'in principle' proteins were molecules like the others; in practice they did not behave like that at all. The chemical characteristics discovered by Fischer and by many others were certainly part of the picture, but many felt that they could not be the whole picture.

The idea that was prevalent at the time, and which held the field for at least a century, from the 1850s to the 1950s, was the theory of the *dynamic state*. As the living cell is in a constant state of activity, so are the proteins: they are breaking down and growing again in a continuous exchange of energy and atoms with their environment. In this framework, the most essential property of the proteins could well be neither the order of the aminoacids nor the three-dimensional structure, but something else, either some unknown chemical property or something as elusive as a *vital force*.

The fact that proteins could crystallize, for example, was not incompatible with the dynamic state: the proteins could well have a variety of different structures *in vivo*, and the crystals could grow from one or a few forms that the proteins assume when extracted from their living environment and transported into a totally artificial medium.

The only thing about which there was general agreement, was the importance of proteins for life. The number of cases where a life-specific chemical turned out to be a protein was constantly increasing, and the variety of jobs that proteins could perform seemed to have no limit. There is actomyosin in muscles, collagen in tendons and cartilage, crystallins in eyes, keratins in skin, fur, hair, wool, nails, horns, scales, beaks and feathers, hemoglobin in blood, albumins in eggs, and lactoglobulin in milk. The antibodies of the immune system are proteins, and so are many hormones secreted by the endocrine glands.

Finally, all enzymes are proteins, and this really makes proteins the staff of life. At any moment, thousands of chemical reactions are going on within the cell to transform and utilize matter and energy. To reproduce these reactions in the laboratory we need long periods of time, high temperatures, low pressures, strong acids and bases and powerful organic solvents, while the cell performs them in split seconds, at moderate temperatures and in nearly neutral solutions based on water. The secret of life lies in the catalytic power of the enzymes, in their ability to accelerate chemical reactions without being affected by them.

In 1894, Emil Fischer proposed that an enzyme interacts with a specific target, or substrate, because they have complementary shapes which allow the substrate to fit into the enzyme like a key fits into a lock. This famous analogy turned out to be even more correct then Fischer imagined. Substrate and enzyme not only have complementary shapes but also complementary electric charges which ensure such a tight fit that no water remains between them. As Max Perutz said:

> There is a profound chemical purpose in this. In water many reactions proceed slowly because of its high dielectric constant: water acts as an insulator which keeps charged molecules apart. The interior of enzymes, on the other hand, is made up of hydrocarbons, which have a low dielectric constant. In this environment, strong electrical forces can be brought to bear on the reactants, causing them to be altered in a small fraction of a second. So enzymes may be regarded as the organic solvents of the cell.

NUCLEIC ACIDS

In 1944, Oswald Avery, Colin Macleod and Maclyn McCarty announced that the transformation of pneumonia bacteria from a mild into a virulent form (a phenomenon discovered by Fred Griffith in 1928), is caused by almost pure deoxyribonucleic acid (DNA). The experiment took biologists by surprise because the idea that DNA is the molecule of heredity had been considered since the beginning of the century, and had been rejected unanimously by generations of scientists.

In 1900, for example, Edmund Wilson wrote that DNA is "the formative center ... the nuclear control of the cell", but later he changed his mind. Chemists convinced him and apparently all other biologists that, as Aaron Levene put it,

> the nucleic acids of the nucleus are, on the whole, remarkably uniform ... in contrast to the proteins which seem to be of an inexhaustible variety. It has been suggested, accordingly, that the differences between cells depend on their protein components and not upon their nucleic acids.

The chemical structure and the biological role of the nucleic acids had been studied since 1869, when Friedrich Miescher extracted from nuclei a substance which contained carbon, hydrogen, oxygen and nitrogen, like proteins, but which was also rich in phosphorus and could not be a protein because it was not digested by pepsin. Miescher called it 'nuclein' but later the substance became known as 'nucleic acid', and was divided in two great families of molecules. Those which contain the sugar 'ribose' are called ribonucleic acids or RNAs; the others contain the sugar 'deoxyribose' (a ribose *de*prived of one *oxy*gen), and are called *deoxy*ribonucleic acids or DNAs.

In a solution of sodium hydroxide, the nucleic acids break down in subunits or building blocks which are called 'nucleotides', molecules which are invariably formed by a phosphate, a sugar and a base. The bases are the identity labels of the nucleotides (like the radicals in the aminoacids), and there are four of them in every family. The bases of the RNAs are A,U,G,C (Adenine, Uracil, Guanine and Cytosine) while those of the DNAs are A,T,G,C (Thymine instead of Uracil).

Every nucleotide carries the groups COH (on the sugar) and HOP (on the phosphate) which creates the bridge COH-HOP between any couple of nucleotides. With the elimination of a molecule of water (H-HO) the sugar-phosphate bridge becomes CO-P which links the two nucleotides together in a strong chemical bond.

These properties show that proteins and nucleic acids have various things in common. They both consist of chains of subunits (aminoacids or nucleotides), and every subunit consists of a backbone and an identity group (a radical or a base). Furthermore, the link between the subunits is obtained in both cases by eliminating a molecule of water from the chemical bridges that join the backbones together.

Despite these similarities, however, chemists found that proteins and nucleic acids behave in totally different ways. While the proteins display an "inexhaustible variety" of properties the DNAs, as Levene said, are indeed "remarkably uniform", they have the same boring and repetitive chemical characteristics in all organisms, very much like starch and cellulose.

To account for this experimental fact, chemists proposed that the real subunits of the nucleic acids are groups of four nucleotides (the tetranucleotide hypothesis), and that the final molecules are monotonous repetitions of tetranucleotides just like starch and cellulose are monotonous repetitions of sugars. These repetitive molecules are the same whatever organism they come from, and cannot therefore be responsible for the hereditary characteristics which differentiate groups of organisms from one another, and each individual creature within a group.

In addition to this chemical reason, there was also a more general argument against the idea that DNA could have an hereditary role. Heredity must transfer the information to build specific proteins, and since the function of a protein is determined by its three-dimensional structure, it was concluded that heredity must transfer *three-dimensional* information. This could have been, for example, the information contained in the form of the hereditary molecules, form which could act as a mold for the reproduction of complementary forms. Alternatively, it could have been the transfer of specific enzymes which catalyze the reactions necessary to perform three-dimensional operations. In either case the carriers of heredity would have to be proteins, because no other substance can match them in building specific forms in space, or in performing elaborate enzymatic reactions.

Avery was aware that his experiments required a new chemical theory of the nucleic acids, and said so explicitly at the end of his paper:

If the results of the present study on the nature of the transforming principle are confirmed, then nucleic acids must be regarded as possessing biological specificity the chemical basis of which is as yet undetermined.

This conclusion however had to face a serious objection: if the DNAs have biological specificity, why is their chemical behaviour so uniform? The tetranucleotide model could be wrong, but the chemical uniformity of the DNAs was a fact, a laboratory fact of which every biochemist had direct experience and which was beyond any reasonable doubt.

THE DOUBLE HELIX

Avery's paper was received with skepticism and disbelief, but a few pioneers followed him, and within a decade the DNA revolution became a triumph. The turning point came in 1953, when James Watson and Francis Crick proposed the double helix, the model that in one swift stroke gave a new chemical theory to DNA and a new mechanism to heredity.

The model was based on four main assumptions. The first hypothesis is that DNA is made of two chains of nucleotides linked together base by base. One can easily visualize it as a ladder, where the uprights are the sugar-phosphate backbones of the two chains, and the rungs are the pairs of bases which hold the chains together. The final image of the double helix is obtained by twisting the ladder around its long axis, making a complete turn every ten rungs.

The second hypothesis is the rule that Adenine pairs with Thymine (and viceversa), and Guanine with Cytosine (and viceversa), so that the order of the bases in one chain automatically determines the order in the other chain.

The third hypothesis is that the chemical bonds between the bases can be broken much more easily than the bonds between sugars and phosphates in the backbones, so that the double helix can easily be 'unzipped' into its two chains of nucleotides. New nucleotides can then be attached to these single chains according to the pairing rules, and two identical copies of the original double helix are obtained. Watson and Crick proposed that this is how genes are split in complementary halves, and how each half reconstitutes the other. A model on the structure of DNA became a model on a function of the cell, a mechanism for the replication of genetic material.

The fourth hypothesis is that the pairs AT, TA, GC and CG can be arranged in any order within the double helix. This does not affect the physical dimension of the molecule because A is larger than T as G is larger than C, and the pairs have very similar lengths; as a result, all double helices have a constant diameter of 20 Angstroms. Furthermore, the order of the base-pairs does not affect the chemical properties of the molecule, because the bases are inside and the external surface of the molecule presents to the environment a uniform repetition of sugars and phosphates. This explains immediately why the chemical behaviour of all DNAs is so uniform; their acidity, for example, is due to the phosphates and cannot be expected to vary.

As three-dimensional molecules, all genes are helices which split, replicate, interact with the environment and aggregate with each other in the same way. Looking for physico-chemical differences among genes is like trying to understand the difference between various books by determining the chemical composition of their pages, instead of reading them.

All this means that heredity cannot be the transfer of 'three-dimensional' information. If Avery is right and if DNA is a double helix, heredity can only be the transfer of *linear* information, because double helices can only differ in the linear sequence of their bases.

THE CHROMOSOME THEORY

The elegance of the double helix was perhaps more effective than experimental evidence. While the discovery of Avery circulated very slowly and was bitterly resisted, the simple beauty of the double helix convinced biologists, almost overnight, that DNA is indeed the molecule of heredity. It should be appreciated however that what gave to the model of Watson and Crick and to the experiment of Avery their full biological meaning, was a wider theory.

After all, Avery proved that DNA is the molecule of heredity in one single strain of bacteria. From this to say that the genes of countless other species are made of DNA seems a long way to go, but biologists knew, for a very special reason, that the meaning of his experiment was precisely that. The special reason was the Chromosome Theory of Heredity, the doctrine that hereditary characters are carried in supramolecular structures called chromosomes.

In eukaryotes, the chromosomes become visible under the microscope during cell division (mitosis), when they condense in rod-shaped particles whose number is a fixed characteristic of every species, and whose chemical composition is a mixture of proteins and DNA.

In prokaryotes, these particles are never seen under the optical microscope, and the cells divide by simple splitting, or fission, not by the complicated process of mitosis. Mixtures of proteins and DNA however could be extracted also from bacteria, and many insisted on calling them chromosomes. Eventually, however, it was shown that prokaryotic chromosomes are made of pure DNA; the proteins which are attached to them represent cytoplasmic contaminations and not structural components like those of the eukaryotic chromosomes.

There is therefore a difference in structure, and not just in size, between the carriers of genes in prokaryotes and eukaryotes, but the Chromosome theory can be given a formulation which is valid for both types: the division of the cell is always preceeded by the separation of the chromosomes into two equal sets. (The sexual cells of the eukaryotes have a third type of division — called meiosis — where the number of chromosomes is halved. With the union of a sperm and an egg, two single — or haploid — sets of chromosomes become again a double — or diploid — set, and the development of a new creature begins).

The extension of the Chromosome theory to all cells is only the latest in a long series of refinements and reformulations which have taken place since the theory started circulating in the 1860s. The theory itself is the result of so many contributions that we cannot attribute it to anyone in particular: it is truly one of the most collective enterprises in Biology. Most important, the theory is based on so many observations in animals, plants and microorganisms that biologists have no doubt about its universal value, and it was this that gave to the experiment of Avery and to the model of Watson and Crick their far-reaching implications.

Even in this wider framework, however, there was still something missing before one could say that the problem of heredity had been solved 'in principle'. Heredity could be reduced to the transfer of linear information, if and only if one could show (A) that only linear information is required to build three-dimensional proteins, and (B) that the three-dimensional structure of a protein accounts completely for its function.

THE STRUCTURAL PRINCIPLE

In 1953 Frederick Sanger determined the order of the aminoacids (or primary structure) of insulin, and soon afterwards the same technique proved that many other proteins have specific sequences. In 1951 Linus Pauling had shown, from X-ray diffractions, that a chain of aminoacids tends to twist itself up, forming what he called an α-helix, also known as the secondary structure of a protein. In 1958 John Kendrew published the first three-dimensional form — or tertiary structure — of myoglobin (reconstructed from X-ray diffraction with a resolution of 6 angstrom), and in 1960 Max Perutz announced the three-dimensional reconstruction of hemoglobin with 5.5 angstrom resolution.

Sanger, Pauling, Kendrew and Perutz proved in this way that

the theory of Emil Fischer is correct: a protein does have a specific sequence of aminoacids and a specific three-dimensional structure. But all this was true *in vitro*, for proteins which had been extracted, purified and crystallized outside the cell, in a totally foreign environment. It was still possible to think that *in vivo* they undergo a variety of structural changes, according to the century-old theory of the dynamic state.

The contribution which effectively dismantled this theory was made by Jacques Monod and his colleagues in a series of papers which appeared between 1952 and 1962. They proved that new aminoacids are not added to old pieces of proteins, and that there is no regeneration of proteins in the cell, only total synthesis from individual aminoacids. Furthermore, they proved that various enzymes of the bacterium E.coli are fully stable in vivo, under normal conditions of growth. The chain of proof culminated with the discovery of the 'allosteric' proteins in 1962. These are enzymes where the attachment of a substrate induces a change of shape that causes the total arrest of enzymatic activity: it was a direct demonstration that the function of a protein depends exclusively upon its structure.

Other experiments proved that a chain of aminoacids folds itself up in space and assumes 'spontaneously' a specific three-dimensional structure. In 1961, for example, Anfinsen and Haber showed that an enzyme can unfold itself completely in a concentrated solution of urea and that, after removing the urea, it refolds itself up and reacquires in full its enzymatic activity.

The experiments of Monod, Anfinsen and many others provided the evidence which buried for good the theory of the dynamic state, and put in its place one of the most fundamental principles of Molecular Biology: *The primary structure is solely responsible for the three-dimensional structure of a protein; the three-dimensional structure is solely responsible for the biological activity of the protein.*

This 'Structural Principle' put at the center of Biology the problem of protein synthesis: how does the cell use the linear information of the genes to build proteins? A chain of nucleotides must be translated into a chain of aminoacids, and this means two things: there must be a *code* which establishes a correspondence between nucleotides and aminoacids, and there must be a *decoder* which actually performs the translation.

THE CENTRAL DOGMA

If the instructions to build proteins are carried by genes, one could expect that proteins are formed in direct contact with the genes, but they are not: they are assembled in the cytoplasm, on particles called *ribosomes*. This simple but fundamental truth emerged from a vast number of experiments performed before 1958 on many organisms, where scientists found submicroscopic particles that were given different names like 'small granules', 'ribonucleoprotein particles', 'macromolecular particles', 'dense granules' and many others.

In 1958 Roberts coined the name 'ribosomes' for these particles, and the conclusion that 'the ribosome is the decoder of genetic information' finally made its formal entry into Biology as a general concept, valid without exception for all cells. This concluded a saga which had started 25 years earlier.

Between 1933 and 1943 Jean Brachet, Torbjorn Caspersson, Arturo Ceruti and a few others published the results of various experiments which were all pointing in one direction:

Organs which synthesize large amounts of proteins are always rich in RNA, which is localized in the nucleolus and in the cytoplasm. All other cells and tissues have a much lower RNA content and much less conspicuous nucleoli. This correlation implies that RNA must play a major role in protein synthesis.

Having reached this conclusion, Brachet went further: he broke open the cells, separated nucleus from cytoplasm, and found that small particles containing proteins and RNA could be isolated by centrifugation from the cytoplasm. At roughly the same time, between 1940 and 1943, similar experiments were performed independently by Albert Claude on animals and by Salvador Luria on bacteria. Claude and Luria were actually looking for infectious agents and viruses, but found that 'small granules' containing proteins and RNA were present in both infected and non-infected cells.

The first explicit correlation between 'ribonucleoprotein granules' and protein synthesis, was proposed by Brachet in 1946 at an international symposium held in Cambridge:

When ribonucleoprotein granules are isolated from red blood cells, they are found to contain a small amount of hemoglobin which cannot be eliminated by repeated washings. In the same way, pancreatic granules contain insulin. These facts point towards the following conclusion: ribonucleoprotein granules might well be the agents of protein synthesis in the cell.

Although no conclusive proof was available, at that symposium Brachet built a strong case for the idea that "DNA is primarily confined to the nucleus, while RNA is mainly found in the cytoplasm, and protein synthesis is associated with RNA". Six months later, in 1947, André Boivin and Roger Vendrely submitted a paper to 'Experientia' in which they carried the scheme of Brachet to its logical conclusion: "DNA makes RNA makes Proteins". Later, this became known as *the Central Dogma* of Molecular Biology.

Like Avery's discovery of 1944, the Central Dogma of 1947 was not widely circulated, which is understandable because the evidence was too limited. But the idea was born, and the tests started accumulating. Within ten years it became established beyond any reasonable doubt that protein synthesis invariably takes place on ribosomes in all cells, prokaryotic or eukaryotic, and that ribosomes do not carry DNA.

As a result, when Francis Crick reproposed the Central Dogma, in 1958, biologists were only too prepared to accept it as a universal rule. Proteins are not made on genes: there must be an intermediary between them, and the intermediary must be RNA. DNA makes RNA (transcription), and RNA makes proteins (translation).

THE MESSENGER

Up to 1961, there was no controversy about which RNA was carrying the instructions for protein synthesis. The ribosomes contain 85% or more of the total RNA of the cell, and it seemed obvious that it was ribosomal-RNA which carried copies of the genes.

In 1961, however, François Jacob and Jacques Monod showed that in E.coli the RNA-copies of the genes (which they called messenger-RNAs) are unstable, while the ribosomal-RNAs are not. Most important, they showed that in E.coli the regulation of protein synthesis (the mechanism that determines which proteins are made in which quantity and when), operates directly on the genes.

When a protein is needed, the corresponding gene is switched on, transcription takes place, messengers are translated and a small number of proteins is synthesized. The number is small because the messenger decays rapidly, and it is this characteristic which is at the heart of the control mechanism of the bacterium. If the messenger was stable, the ribosomes of E.coli would keep on synthesizing proteins that the cell may not need any longer. A short-lived messenger, instead, ensures that only a small amount is produced, and if the cell needs more it simply has to keep switching on the corresponding gene. With a regulation mechanism based on short-lived messengers, the bacterium is able to control the synthesis of its proteins to a very high standard of fine tuning, making only the types that are needed and when they are needed.

This however is not the only existing mechanism in Nature, and the cell can achieve a fine control of protein synthesis even with long-lived messengers, for example by regulating the activity of the ribosomes. In this case, the ribosomes could carry stable messengers, but would synthesize proteins only for as long as they are instructed to do so. This is an example of systems which control protein synthesis at the level of translation, in the cytoplasm, and which are called for this reason '*cytoplasmic control mechanisms*', while the alternative systems which operate at the level of transcription are called '*genetic control mechanisms*'. Clearly, a genetic control requires short-lived messengers, while a cytoplasmic control needs stable or long-lived messengers.

Jacob and Monod discovered the genetic control mechanisms in E.coli, but did not generalize their results to all cells because they were well aware that in some eukaryotes (for example in the alga Acetabularia) there is strong evidence for cytoplasmic controls. Their thesis, instead, consisted of two main points:

(A) The genetic control mechanisms are a reality, in the sense that at least some prokaryotes do use them to control protein synthesis.

(B) The RNA-copies of the genes, the messengers, can be short-lived in some cases and long-lived in others, but in all cases they form a distinct class of molecules, and have nothing to do with the RNAs of the ribosomes.

Both points were tested with a variety of experiments and found to be in total agreement with the evidence. Messengers could even be 'seen' at the electron microscope (with some interpreting of the pictures), and were quickly put in a class of their own, even if they represent only a tiny fraction (2% or less) of the total RNA-content of the cell. We have therefore at least two distinct partners in the translation business: ribosomes and messengers.

CODE, ADAPTORS AND FACTORS

If a chain of nucleotides is translated into a chain of aminoacids, there must be a 'code of correspondence' between them; and since there are only 4 nucleotides to code for 20 aminoacids, any aminoacid must be specified by a group — or a 'codon' — of nucleotides. With groups of 2 nucleotides there are only 16 different combinations, while groups of 3 nucleotides allow 64 combinations (4^3). In order to account for all existing aminoacids, therefore, the codons must have at least three nucleotides.

This is the essence of the proposal made by Alexander Dounce in 1952 and by George Gamow in 1954: there exists a code of correspondence between genes and proteins, and every aminoacid is specified by a codon of three nucleotides. The correspondence has become known as the 'Genetic Code' ever since, but it should be noticed that genes do not take part in it. The code is implemented on the ribosomes, between the nucleotides of the messenger-RNAs and the aminoacids of the nascent proteins.

Dounce was aware that the interaction between aminoacids and nucleotides required various specific catalysts and predicted the existence of proteins that later were discovered; today we call them 'activating enzymes'. In 1957 Francis Crick added that there must be yet another type of intermediary molecules, and predicted that these would be new types of RNAs, capable of recognizing a codon by a complementary sequence of three nucleotides that he called 'anticodon'. He named these molecules 'adaptors', but soon afterwards they were discovered by

Hoagland, Zamecnik and Stephenson, and became known as 'transfer-RNAs'.

The correspondence between aminoacids and nucleotides is established in the following way: an aminoacid is attached to a specific transfer-RNA by an activating enzyme, and then the transfer-RNA is attached to a triplet of nucleotides on the messenger by a codon-anticodon link.

In addition to that, the synthesis of a protein requires other batteries of molecules. One family of enzymes intervenes at the various stages of translation (they are called initiation, elongation and termination factors), and another class is in charge of the preparation and the maintenance of the transfer-RNAs. We have therefore a variety of molecular systems in translation: ribosomes, messengers, transfers, activating enzymes, maintenance enzymes and factors. In a typical bacterium like E.coli, more than 30% of the cell proteins and more than 70% of the nucleic acids are devoted exclusively to protein synthesis: the translation machinery is the most complex, the most demanding and the most sophisticated single-purpose apparatus that Nature has ever created.

In 1961, at the Fifth International Congress of Biochemistry in Moscow, Marshall Nirenberg announced that protein synthesis can be carried out in vitro with artificial messengers, and his classic experiment opened two new fields of research. On the one hand, the in vitro systems allowed one to determine the exact components which are essential in protein synthesis and to study in detail the various steps of the reaction. On the other hand, Niremberg showed that a nucleotide chain containing only uracil was translated in an aminoacid chain containing only phenylalanine: he had deciphered the first letter of the genetic alphabet. Two years later, thanks to the work of hundreds of scientists, the Genetic Code was completely worked out.

RIBOSOMES

Many protagonists of protein synthesis — aminoacids, messengers, transfers and activating enzymes — are free diffusing molecules in the cell. What brings them together at short distances and in a specific order are the ribosomes, the molecular machines where messengers are decoded and proteins are assembled.

For this function the ribosomes must be conveniently large, and from a molecular point of view in fact, they are superstructures. Their molecular weights range between 2 and 3 million in prokaryotes, and between 3.5 and 4.5 million in eukaryotes; at the electron microscope, where they look like spheroidal particles, their diameters range between 200 and 300 angstrom, the dimensions of small viruses. Having the size, the weight and the composition of small RNA-viruses, the ribosomes were also expected to have the same degree of complexity, in the 1950s, and were often described as virus-like particles, but things turned out to be very different.

In 1961 Waller and Harris analyzed the ribosomes of E.coli by one-dimensional gel electrophoresis, and discovered that there are at least 20 different proteins in them. In 1970 Kaltschmidt and Wittmann introduced two-dimensional electrophoresis in the field, and found many more. This technique revealed that there are between 50 and 60 different proteins in prokaryotic ribosomes, and between 70 and 80 in eukaryotic ribosomes. Furthermore, Ira Wool showed that ribosomal proteins and ribosomal nucleic acids are consistently larger in eukaryotes than in prokaryotes, a discrepancy which "is a paradox since eukaryotic ribosomes perform the same general function".

The meaning of these data can be appreciated by comparing viruses with ribosomes. Viruses contain nucleic acids (DNA or RNA) surrounded by a protein coat, or shell, which is made from very few types of proteins. Most viruses have only one type, and those which have more normally do not exceed 4 or 5 kinds of different proteins. We realize in this way that ribosomes are between 10 and 20 times more complex than viruses, in terms of number of different components and three-dimensional structures.

In the 1950s, as we have seen, the technology to reconstruct the structure of macromolecules from X-ray diffractions of their crystals became available, and has been applied to many proteins ever since. The same procedure could have been used with well-known crystals of viruses, but the sheer complexity of the project delayed its implementation for nearly 20 years. The technological problems of virus crystallography were solved at the end of the 1970s, and the first models appeared in 1980.

In the case of the ribosomes there is a complexity gap between them and the viruses which is similar to the gap between viruses and proteins, but on top of that there is another obstacle in the way: at present we do not have crystals of ribosomes as good as those of proteins and viruses. Some progress in this direction, however, has been made. The first report that true mature ribosomes can form small crystals within the cell came in 1966, when Gianfranco Ghiara and Carlo Taddei found them in lizard oocytes and Breck Byers induced them in chick embryos by cooling. In 1970 my colleagues and I managed to take these microcrystals out of the cells, and in 1979 I crystallized eukaryotic ribosomes in homogenates and cell extracts of chick embryos; that was the first case of eukaryotic ribosome crystallization outside the cell (Figure 6). Three years later, Wittmann and colleagues (1982) announced the crystallization in *in vitro* of prokaryotic ribosomes.

Let us now look at the general meaning of these reports. Structures like nucleoli, mitochondria or chloroplasts are more complex than ribosomes, but are so varied in size and shapes that no two particles are identical, and crystallization is impossible. They are *supramolecular* structures, while ribosomes maintain, like all true molecular systems, the *potential* to crystallize. We realize in this way that ribosomes have a unique place in Nature, at the very top of the hierarchy formed by all molecular systems: there simply is no other object in the known Universe that can form crystals and has a higher, or even a comparable, degree of complexity.

This makes us appreciate what protein synthesis really is: in order to carry it out, Nature had to invent the most sophisticated molecular machine that has ever been assembled. The ribosomes are its crown jewels, the ultimate result of all the molecular engineering that Nature has put into life.

REVERSE FLOWS OF INFORMATION

In 1964 Howard Temin discovered a link between the malignant transformation induced in some eukaryotic cells by RNA-viruses and an enzyme that makes DNA from RNA. At that time the link was not fully proved, but six years later Temin himself,

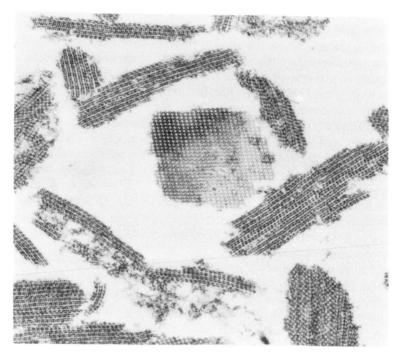

FIGURE 6:   Ribosome microcrystals obtained in cell extracts of chick embryos (Barbieri, 1979).

David Baltimore and Sol Spiegelman (1970) were able to provide conclusive evidence for it.

Since the process by which 'RNA makes DNA' is the opposite of transcription, it has become known as 'reverse transcription' and the enzyme that promotes it as 'reverse transcriptase'. In 1970 Francis Crick pointed out that the new process modifies but does not violate the Central Dogma, which could now be written in the following way:

$$\overset{\curvearrowleft}{DNA} \rightleftarrows RNA \rightarrow Protein$$

The essence of the new version is that all types of information transfer initiated by nucleic acids are possible, while all those initiated by proteins are forbidden. We still do not know if reverse transcription is used predominantly by RNA-tumor viruses or is a much more general phenomenon. So far, it does not seem to take place in bacteria but, as Temin suggested, it could operate in normal eukaryotic cells and could be a very important mechanism of differentiation, of immune-response and of evolution.

Some have gone further. If reverse transcription is possible, why not reverse translation? Mekler (1967) and Cook (1977) have shown that there is nothing wrong with reverse translation from a theoretical and from a biochemical point of view. There is however a real obstacle in the way. While reverse transcription needs only a protein that does exist, reverse translation would require *reverse-ribosomes* which, as far as we know, are nowhere. One can speculate that reverse-ribosomes might have existed in the past, or that only an accident prevented their evolution, but the fact remains that the translation of proteins into RNA would need a completely different machinery from the translation apparatus of the cell, and that machinery simply does not seem to exist.

This is why most biologists accept the version of the Central Dogma proposed by Crick in 1970, and believe that we can write off the inheritance of acquired characteristics for good, but some still have doubts. Ted Steele, for example, has recently proposed that the immune system is a special case, possibly an important exception (1979). Even if he turns out to be right, however, there is little chance that the inheritance of acquired characteristics will ever become a major mechanism of evolution. The existence of ribosomes in all cells and the non-existence of reverse-ribosomes tell us that the flow of biological information is from nucleic acids to proteins, not the other way round. That is the solid foundation of the Central Dogma, and that is a very well documented fact of life.

SELFISH AND SILENT DNA

Of all the surprises that have come from Biology in recent years, two have been particularly effective in giving "a good shake" (as Crick put it) to our ideas. Not so long ago biologists were taking almost for granted that all DNA is 'useful' and that there is 'colinearity' between genes and proteins ('useful DNA' includes the 'structural genes' that code for proteins and the 'control genes' that regulate the expression of other genes. 'Colinearity' means that the number and the order of codons in a gene is equal to the number and the order of amino-acids in the corresponding protein).

As a result, it was generally believed that more complex organisms would simply have more DNA than lower creatures, i.e. that the amount of DNA per cell is a straightforward measure of biological complexity. In practice, however, the assumptions that all or most genes are useful and that they are colinear with proteins, turned out to be valid only in bacteria. In eukaryotic cells of all phyla, the majority of DNA is 'nonsense', and there is no relationship between number of genes and biological complexity. There are certain salamanders and toads, for example, that have up to 50 times more DNA per cell than humans do.

There are two distinct types of 'nonsense DNA'. The first is formed by sequences that make no contribution to the phenotype and that are found 'outside' the useful genes. This DNA, discovered by Roy Britten in 1968, is often formed by the repetition of simple sequences, and Britten called it 'repetitive DNA'. Later on, Dawkins (1976) and Orgel and Crick (1980) labelled it 'selfish DNA'.

The second type of nonsense DNA differs from the first because normally it is not repetitive, and in particular because its sequences are found 'inside' the useful genes, which are in this way 'split' into pieces. More precisely, many useful genes of eukaryotic organisms contain nonsense pieces of DNA interspersed with pieces that make sense. In these cases, the cell produces a full RNA transcript of the DNA, and then cuts out with 'splicing enzymes' the nonsense sequences before sending the messenger to the cytoplasm (the discovery of split genes and RNA-splicing was announced simultaneously by many authors at the Cold Spring Harbor Symposium on Quantitative Biology in 1977). Gilbert (1978) gave the name of 'exons' to the pieces of DNA which are actually expressed in messengers, and of 'introns' to the unexpressed ones, which also became known as 'silent DNA'.

Some biologists are not convinced that selfish and silent DNAs deserve the qualification of nonsense (or junk) DNA, because the cell may well use them for some unknown useful purpose. The names however have become familiar, and we may keep using them as convenient labels. What is certain is that nonsense DNA is not expressed, and that it accounts for the largest share of DNA in most if not all eukaryotes. The DNA of human chromosomes, for example, could contain more than 2 million genes, while the highest estimates for the genes that are actually expressed do not exceed 100,000.

The origin of huge quantities of unused DNA is clearly a major evolutionary problem. On average, eukaryotes have a thousand times more DNA per cell than prokaryotes, but even the most complex animals have only between 10 and 30 times more useful genes than bacteria. E.Coli, for example, has about 3,000 structural genes, while in Drosophila they are less than 10,000, in mice around 30,000 and in man no more than 100,000.

Some have suggested that complexity should be related not to DNA content but to differences in DNA sequences, but the proposal does not seem to be meaningful. There are two species of Drosophila flies (*D.melanogaster* and *D.funebris*) which look very much alike, when in fact they have only 25% of DNA sequences in common. By contrast, cows and sheep share 89%, while humans and chimpanzees have in common 98% of their DNA sequences.

Another suggestion was that animals which appeared earlier in evolution should have less DNA than later creatures, but this turned out to be wrong; many fishes, for example, carry far more DNA than mammals. The history of life, in other words, was not accompanied by a steady increase in DNA content. We can definitely say that eukaryotes carry more DNA than prokaryotes, but apart from this there is no other regularity in the evolutionary pattern of DNA.

And yet there must be an explanation. We cannot say that nonsense DNA was impossible to avoid, because prokaryotes managed to do just that. And we cannot say that it came by accident, because the phenomenon is too generalized and affects virtually all eukaryotes. But what was the advantage of carrying vast quantities of DNA if the cell had then to produce enzymes to get rid of their transcripts?

In 1978, Gilbert, Doolittle and Darnell made a very interesting suggestion. Different splicing enzymes would allow different pieces of RNA to be cut out and create different messengers from the *same* gene. In this way the cell would not have to wait for chance mutations to experiment with new proteins; it could directly produce a large number of different proteins simply by trial and error experiments with splicing enzymes. For Doolittle and Darnell this mechanism of 'RNA shuffling' is so advantageous that they believe it originated at a very early stage in the history of life.

If Gilbert, Doolittle and Darnell are right (personally I believe that they are), the eukaryotes had a very powerful mechanism to generate biological diversity. But how did the mechanism originate in the first place? And why did the prokaryotes not exploit it if it was so advantageous? These are questions that go to the very heart of the strategy of evolution, and later we will have to discuss them carefully. For the time being, let us say that the amount of DNA *per se* did not mean much. It is becoming increasingly clear that Pantin (1951), Britten and Davidson (1971), Zuckerkandl (1976) and Løvtrup (1977) were right when they said:

evolution is not a question of making new proteins, but rather of using old proteins for new purposes.

NEW AND OLD

Molecular Biology has changed the life sciences. Concepts like the double helix, the genetic code, the central dogma, ribosomes, messengers, transfers, transcription and translation and many more, simply did not exist before the 1950s. And yet the two fundamental concepts of Classical Biology have not changed. The cell is still a duality of genotype and phenotype, and correspondingly the mechanism of evolution is still a two-step process of variation in the genotypes and selection in the phenotypes.

Some may take comfort in this, but of course the continuity of ideas is desirable only when they give us a satisfactory framework, and in evolution this is not the case. The answer of the Classical theory is still the same, but so are the objections to it, and the doubts. The controversies of today about the history of life are not substantially different from those that were raging 50 or 100 years ago. Why? Is it because we have no substantial new evidence, or because we keep looking at the evidence from a theoretical point of view that despite all the novelties remains fundamentally the same? This is what really matters today in Biology, and it is this problem that we will try to discuss in the rest of the book.

CHAPTER 5

Theories of Precellular Evolution

FIVE THEORIES

The idea that the first microorganisms evolved from lower forms by a long process of chemical developments started circulating in the 1860s, in the climate created by the experiments of Pasteur on spontaneous generation, and by the publication of the Origin of Species.

The idea of 'Precellular Evolution' is the only conceivable alternative to the idea of 'Sudden Life', and in this sense it is one concept. We have seen however that there are five very different theories of Sudden Life, and we should keep in mind that (so far) there are also five different theories of Precellular Evolution. These theories differ from each other (in principle and not just in technical details) because they attribute the critical role to different substances. These are:

1) Proteins   (the Phenotype theory)
2) Genes   (the Genotype theory)
3) Viruses   (the Viral theory)
4) Minerals   (the Mineral theory)
5) Ribosoids   (the Ribotype theory)

In this Chapter I will say briefly why I believe that the first four theories are unsatisfactory, whereas the term ribosoids and the Ribotype theory will be discussed in Chapter 7.

THE PHENOTYPE AND GENOTYPE THEORIES

The first theory of precellular evolution was proposed by Alexander Oparin in 1924, with the suggestion that primitive cells evolved from coacervates of proteins. At that time, proteins were thought to be capable of continuous metabolism (the theory of the 'dynamic state'), and were regarded as being as alive as cells are. Furthermore, Johannsen's distinction between genotype and phenotype was seen as a mere division of labour among proteins, not as a duality of material substrates. Proteins were thought to account not only for metabolism but also for heredity, and in this framework nothing else but proteins could have created the first cells by natural causes. Oparin's theory therefore was bold and revolutionary, but was also very logical; from an evolutionary point of view it was the only solution to the origin of life that was consistent with the biochemistry of the time.

All this changed drastically when it turned out that hereditary characters are carried by nucleic acids. Now there were two distinct types of molecules at the basis of life: proteins for metabolism and nucleic acids for heredity. As a result, two distinct solutions to the problem of the origins became conceivable: the first cells evolved either from a protein-path (the Phenotype theory) or from a gene-path (the Genotype theory). In 1957 Oparin reproposed the protein-path of his original scheme, and his model represents what I call the Phenotype theory on the Origin of Life, the view that metabolism preceded heredity, that proteins had an evolutionary priority over the nucleic acids.

Many other biologists however chose the Genotype theory, the idea that the first cells evolved from naked genes. This statement, as Doolittle and Sapienza remarked in 1980, "has been made so often that it is hard to remember who made it first" and the paternity of the Genotype theory is uncertain. There is no doubt however that the theory became, and still is, very popular. Statements like "genes created the cell as their throwaway survival machine" and "organisms are DNA's way of producing more DNA", have become familiar expressions.

Despite this popularity, the Genotype theory has never achieved an undisputed supremacy because there is in it, as in the Phenotype solution, a core of irrationality. Both theories lead straight to the classic paradox of the chicken-and-the-egg, where proteins are primordial chicken and genes are primordial eggs. Both sides try to avoid the paradox by saying that after an initial stage *proteins and genes evolved in parallel*, but this is precisely where the paradox comes in. Coacervates of proteins containing random nucleic acids, or clusters of genes attached to random proteins, are conceivable, but the parallel evolution of genes and proteins towards an integrated unity is a much tougher proposition.

Coacervates of proteins, for example, could have exploited nucleic acids if information could be transferred from proteins to genes (by some kind of reverse translation), but this is the opposite of what Molecular Biology has discovered for the present, and we have no plausible model which accounts for such a mechanism in the past. On the other hand, in a cluster of proteins made from DNA information would flow in the right direction, from genes to proteins, but this is not enough to avoid the paradox. Consider for example how genes could have surrounded themselves with membranes. They could have found membrane proteins already existing around them, but the replication of these proteins would have required a transfer of information from proteins to genes. Alternatively, a gene could have appeared by chance with precisely the right sequence to code for a membrane protein, but in order to synthesize it with even a low degree of accuracy it would have needed the translation apparatus of a primitive cell. A cell would have to exist in order to build the components which make up the cell in the first place.

The statement that "proteins and genes evolved in parallel" therefore is no magic which melts the paradox away. Unless a consistent mechanism is found for it, it remains wishful thinking. The supporters of the Phenotype and Genotype theories are confident that some day a suitable mechanism will be found, and this is undoubtedly a legitimate expectation. We have seen however in the history of science that sometimes the solution of paradoxes has required a total departure from the theories that created them, and there is no guarantee that this will not happen in our case. For the time being therefore the Phenotype and the Genotype theories remain inherently paradoxical. For the future we will just have to wait and see.

THE VIRAL THEORY

The idea that viruses played a key role in precellular evolution has a fair number of supporters, and was illustrated by Efraim Racker in the following way:

A molecular biological version of the origin of life runs something like this. On the first day God said, "let there be light", and there was light. On the second day God said, "let there be water", and there was water. On the third day God said, "let there be membranes which contain chlorophyll and which can utilize the energy of light". On the fourth day the chlorophyll-containing vesicles made a~, and the squiggle generated ATP, GTP, UTP, and CTP. On the fifth day God polymerized ATP, GTP, UTP and CTP and created ribonucleic acid and God saw that it was good. On the sixth day the Lord said it was not good for RNA to be alone and He caused a deep sleep to fall upon RNA (He incubated at 0°C) and He took a rib out of (rib)onucleic acid and made DNA from it (the rib was later renamed oxygen), and RNA and DNA were both naked but they were not ashamed. On the seventh day God rested and that's when all the trouble started. There was the ~ with its bad influence and persuaded DNA to do what it was not supposed to do. DNA and RNA made hybrids and afterwards they felt ashamed because they were naked. So they covered themselves by making a coat protein and this is how the first virus was created.

God was very angry when He saw it and blamed RNA, and RNA blamed DNA and DNA blamed the ~ (this was the first case of passing the buck). But God did not like to have viruses float around in Eden and He banned RNA and DNA onto Earth where they lived unhappily ever after.

This lively parable contains two interesting ideas. The first is that RNA preceeded DNA in evolution, a concept that is generally accepted even by the supporters of the Genotype theory, because RNA is more reactive and more flexible than DNA. The second is that the most primitive genotype-phenotype system was a virus.

Racker however did not add that the real trouble begins precisely at this point: how do we go from viruses to cells? Here, as in Genesis, the curtain falls with the departure of the progenitors from Eden, and when it rises again we find that the successful population of the Earth has already taken place.

To my knowledge, the Viral theory of precellular evolution was first proposed by J.B.S. Haldane in 1929. A few years earlier, in 1917, Felix d'Hérelle had discovered the bacteriophages, and biologists were passionately discussing if these viruses could or could not be regarded as true living creatures. Whatever side one took, it seemed clear that viruses are at the very border which divides life from non-life, and Haldane concluded that they could well have been the historical precursors of the first cells.

He proposed this idea together with the hypothesis that there was little or no oxygen on the primitive earth, a point that was

D

extended by Harold Urey in 1952 and on which there is today a general consensus (Oparin incorporated this concept in his theory in 1936). Haldane's theory is often associated with Oparin's, but I prefer to keep them apart because Oparin gave an evolutionary priority to metabolism, while Haldane put the emphasis on reproduction. The spirit of his Viral theory is much nearer to the Genotype theory of the naked gene than to the Phenotype theory of Oparin.

Haldane however did not say how viruses evolved into cells, and nobody else has. As we have seen from the account of Efraim Racker, the theory gives us a beautiful description for the origin of the viruses, but then it does not go further and this is a pity because from all that we know viruses need cells to replicate. Perhaps this is why Racker chose a biblical style for his story: we can take it or leave it, but presumably we are not supposed to ask indelicate questions.

THE MINERAL THEORY

The formation of proteins and nucleic acids, as we have seen, requires the elimination of a molecule of water from any couple of aminoacids and of nucleotides. In primitive solutions this means that molecules had to be dehydrated in water, which seems impossible while in fact it is only very difficult. The chemical carbodiimide, for example, reacts more vigorously with the water that is 'potentially' present in certain molecules than with free water itself, and has already been used to form short nucleotide chains in the laboratory. Other condensing agents are cyanamide and dicyanamide, much more likely to be present on the primitive earth, but perhaps the most interesting ones are surface catalysts like silica and clays.

In 'The Physical Basis of Life' (1951) John Bernal suggested that the adsorption to clays, muds and inorganic crystals are powerful means to concentrate and polymerize organic molecules, and the idea has not lost its appeal ever since. Most condensation reactions in the laboratory invariably give better results when the mixture is allowed to dry, which suggests that the evaporation of primitive solutions splashed on rocks, muds and beaches may well have been a very important mechanism of chemical evolution.

In 1966 and 1982 Cairns-Smith took this idea much further. While Bernal proposed that silica, clays and crystals only 'helped' in the formation of organic molecules (a concept that is almost universally accepted), Cairns-Smith suggests that "clays were the materials, perhaps the sole materials, out of which the earliest organisms were *made*".

The theory was vividly summarized in the opening sentence of his 1966 paper:

It is proposed that life on Earth evolved through natural selection from inorganic crystals.

In the preface to 'Genetic takeover and the mineral origins of life' (1982) Cairns-Smith restates the idea with equal concision:

the first organisms on Earth had an altogether different biochemistry from ours — they had a solid-state biochemistry.

The argument is developed in two logical steps. The first part is an all-out attack against the spontaneous formation of polypeptides, polysaccharides and particularly polynucleotides in primitive solutions. Cairns-Smith makes a list of no less than 19 objections to the prebiotic origin of nucleic acids alone, and maintains that the real importance of the simulation experiments

lies not in demonstrating how nucleotides could have formed on the primitive Earth, but in precisely the opposite: these experiments allow us to see ... just why prevital nucleic acids are highly implausible.

In the second part, Cairns-Smith builds his alternative with the concept of *takeover*.

Consider an analogy with human technological development. Primitive ways of doing things — with a clay tablet or a stone axe — depend on very little fabrication. Later ... much more sophisticated means to similar ends became possible ... Perhaps the axe head became a spear, and the spear a bow and arrow, but the intercontinental missile has a quite different history. That sort of deviousness is quite like evoluton too. Our lungs are not improved gills, nor is our way of walking related to the locomotion of an amoeba. During evolution quite new ways of doing things become possible from time to time and these may displace older ways ... It is in this kind of way that evolution has been able to escape from original design approaches — through takeovers: first ways may not only be transformed, they can be *replaced* by later ways that are based on unrelated structures.

Having established this general principle, Cairns-Smith asks the crucial question:

Can we imagine genetic materials that could more easily have been generated on the primitive Earth than nucleic acids?

The answer is yes, and the Mineral theory is developed in four hypotheses that Cairns-Smith states as follows.

1) The most favourable prevital conditions would have been on an Earth that had land and sea and weathering cycles and an atmosphere dominated by nitrogen and carbon dioxide.

2) The first organisms were a subclass of colloidal mineral crystallites forming continuously in open systems.

3) These mineral organisms evolved modes of survival and propagation that would have seemed highly engineered or contrived. That is to say they became a form of life.

4) Some evolved primary organisms started to make organic molecules through photosynthesis. This led to organisms that had both inorganic and organic genes. Eventually the control of their own synthesis passed entirely to the organic genes (nucleic acids) which by now operated through the synthesis of proteins.

Cairns-Smith frankly admits that "genetic takeover is meant to be a background hypothesis. It is meant to be vague — in the sense that it is a general frame within which to make more specific speculations". The main point for him is to reach the conclusion that "there must have been at least one other kind of genetic material before nucleic acids", and the main obstacle in the way is not experimental but theoretical. It is the principle that Nature followed a continuous line of development, the concept that Leslie Orgel expressed in this way:

I shall be guided by a Principle of Continuity which requires that each stage in evolution develops 'continuously' from the previous one.

According to Cairns-Smith there have been periods of continuity interrupted by takeovers during the history of life, and in principle therefore the origin of the nucleic acids could have been either a continuous process or an example of takeover. What makes him decide that it was a takeover is not the principle itself, but the experimental failure of the attempts to produce nucleic acids in vitro in the past 30 years or so. In this situation there is a natural thing to do: let us make simulation experiments with clays and see if we can create inorganic life in vitro. Since the whole theory rests on the hypothesis that mineral life "could have been generated much more easily", we should have the answer in far less than 30 years time.

I do hope that these experiments will be performed, even if I happen to believe that the theory is wrong, because it is always a good precaution to look for evidence that may disprove our own ideas. Personally I believe that the Mineral theory is wrong because it amounts to this: since we have failed to create primitive living systems based on carbon, we must conclude that the first living systems were based on silicon. Now, from what we know in chemistry ad biochemistry it would seem that silicon is far *worse* than carbon as far as life is concerned. If so far we have failed with carbon, therefore, it is unlikely that we will succeed with a worse material, but of course we should try. One never knows.

THE SIMULATION EXPERIMENTS

The three photographs that are reproduced in Figure 7 appeared originally in 'The Mechanism of Life' (1907) by Stephane Le Duc, and seem to represent a group of mushrooms, a colony of algae (Acetabularia), and a cell undergoing mitosis. In fact they are all inorganic artefacts that Le Duc created in saturated solutions of potash with dyes, phosphates, chlorides and other salts.

By changing the concentrations and by adding or subtracting various substances, Le Duc was able to make inorganic aggregates grow, expand, divide and mimic living forms with astonishing virtuosity. In the same year, by using the "miraculous" energy of radium, Martin Kuckuck showed that a mixture of

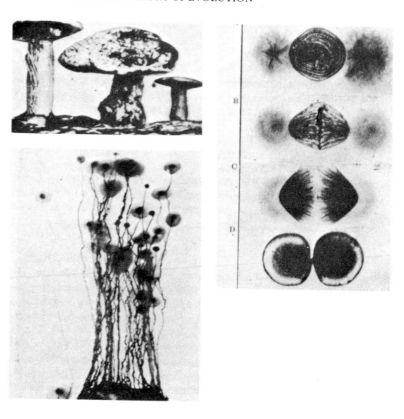

FIGURE 7:   Inorganic artefacts mimicking mushrooms, Acetabularia cells and mitosis (from *The Mechanism of Life* by Stephane Le Duc, 1907).

gelatin, glycerol and salts develops "cells" which dance and multiply and are even capable of engulfing particles like amoebae do. These and similar phenomena came to be regarded as the beginning of a new science that Le Duc called "synthetic biology" and Kuckuck named "la biologie universelle". Synthetic and universal they may have been, but as far as biology is concerned the osmotic creations of Le Duc and Kuckuck were no more alive than soap bubbles.

A really new field of research began instead in 1953, when Stanley Miller obtained aminoacids from a gaseous mixture of methane, ammonia, water and hydrogen exposed to an electric discharge for a week. That classic experiment and others that followed on the same line are today referred to as the 'simulation experiments' in chemical evolution, or prebiotic chemistry. They differ from the synthetic biology of Le Duc because the results are not judged by the ability to mimic living forms, but by the relevance of the underlying chemical reactions to the chemistry of life (the chemical reactions of Le Duc could perhaps be relevant to the mineral life predicted by Cairns-Smith, but have nothing to do with organic chemistry).

Today, owing to the work of Miller himself, Melvin Calvin, Leslie Orgel, Cyril Ponnamperuma, Sidney Fox and many others, prebiotic chemistry has become a science in its own right, but the interpretations are still widely divergent. Cairns-Smith, as we have seen, believes that the experiments have failed in their primary goal and have only a negative value, while Sidney Fox, at the opposite side, maintains that his "proteinoid microspheres" or "protocells" are veritable examples of primitive life in the laboratory.

Various people have serious reservations about Fox's protocells, but this does not mean that the negative judgement of Cairns-Smith is justified. Any genuine simulation experiment is necessarily a small piece of an enormous puzzle, and it is not surprising that progress in this area is painfully slow. There is however another side to consider. Somehow, some primitive systems found a way to split water, and if we can learn how that was done we may be able to obtain oxygen and hydrogen safely and in unlimited amounts. Or we could develop new biotechnologies by learning how the machinery of protein synthesis came into being. Either curiosity for its own sake or the prospect of practical applications, therefore, will probably keep the simulation experiments going for the foreseeable future, whatever the difficulties, the criticisms and the blunders.

EVOLUTION IN LITTLE BAGS

In addition to the five schemes that I have already listed, there is a separate class of theories on the origin of life which give a conceptual priority not to any particular substance, but to heterogeneous mixtures of organic molecules which happened to be trapped by membranes in little vesicles. The formation of membranes, vesicles and various organic compounds in the primitive solutions was likely, and the idea that life started in little bags full of chemicals, therefore, is an attractive proposition. After all, bacteria look precisely like little bags under the optical microscope.

This approach has been actively investigated in recent years, and to my knowledge the best examples are the theories proposed in 1977 by Carl Woese (the Progenote theory) and by Eigen and Schuster (the Hypercycle theory). If little bags of chemicals started life, however, it seems to me that they deserve to be regarded as the most primitive cells, and for this reason I will discuss the corresponding ideas in the next chapter, as theories of cellular evolution.

CHAPTER 6

Theories of Cellular Evolution

SYMBIOSIS AND THE PROKARYOTIC THEORY

In 1866 Ernst Haeckel published the first phylogenetic tree which linked together all forms of life, from bacteria to man (Figure 3). He proposed that life began with bacteria (or monera, as he called them), and that some of their descendants gave origin to nucleated microorganisms (protista), which in turn gave origin to all visible creatures.

He had no paleontological or biochemical evidence for this conclusion, and did not hesitate to invent imaginary creatures to illustrate his case, but biologists felt that he had raised a serious point: bacteria must have come first because they are the simplest living creatures. For anyone who believed in evolution this idea was irresistible, and soon after Haeckel the main point was not 'if' but 'how' did the bacteria evolve into protista.

Before the end of the century there were already two distinct theories. One assumed that the transition took place gradually, with all intermediate steps between the simplest and the most complex cells. This is the theory of 'Direct Filiation', according to which there is only one type of cell in Nature that simply evolved into a continuum of different forms. Later, the intermediate cells disappeared, leaving a huge gap in the middle that creates the impression of two distinct forms of life.

The second theory assumed that protista evolved from bacteria not by the accumulation of small changes but by 'jumps', by the exchange of entire subcellular systems. In the 1880s mitochondria and chloroplasts had barely been discovered when Schimper (in 1883) and Altmann (in 1890) proposed that these organelles (which have the shape and the dimensions of small bacteria) were once free-living cells that started living in symbiosis with others and eventually were engulfed by their hosts.

This is the theory of 'Symbiosis' (also known as 'Hereditary Symbiosis' or 'Endosymbiosis'), according to which the eukaryotic cell is a 'multiple' of the prokaryotic cell, a mixture produced by the coalescence of different bacteria in one working unit. The theory of Symbiosis assumes therefore that there is a fundamental dichotomy between prokaryotes and eukaryotes, in the sense that the first are 'elementary' cells, while the second are 'composite' cells (One could compare this dichotomy with the difference between atoms and molecules: the molecules are new units in their own right because their characteristic properties simply do not exist when the atoms are separated).

The Symbiotic theory had a few supporters, who also refined it (Mereschkovsky in 1905, Portier in 1918, Wallin in 1927, Lederberg in 1952), but up to the 1960s it was generally regarded as too extravagant to be taken seriously. What changed people's minds was the discovery that mitochondria and chloroplast contain molecular systems which are definitely prokaryotic. They have 70S ribosomes, for example, and carry their own DNA.

Lynn Margulis reproposed the Symbiotic theory in 1967 (as Lynn Sagan) and pointed out that the same mechanism can account not only for mitochondria and chloroplasts but also for flagella and microtubuli (from spirochetes). This did not convince everybody, but most biologists agreed that the evidence for symbiosis is very strong.

One distinction however becomes necessary at this point. Up to 1970 the followers of Direct Filiation and Symbiosis shared the common idea that prokaryotes came first and evolved into eukaryotes. They had different mechanisms in mind, but the evolutionary priority of the prokaryotes (Haeckel's theory) was unquestioned. From 1970 onwards however new theories appeared which are compatible with Symbiosis but not with Haeckel's view, and these two ideas therefore must be clearly separated.

All models which accept the evolutionary priority of bacteria can be regarded as different versions of the theory of Haeckel, which I call 'the Prokaryotic theory' of cellular evolution. With the five-kingdoms classification proposed by Whittaker (1959), Whittaker and Margulis (1978) and Margulis (1981), the Prokaryotic theory can be summarized by the scheme that is represented in Figure 8.

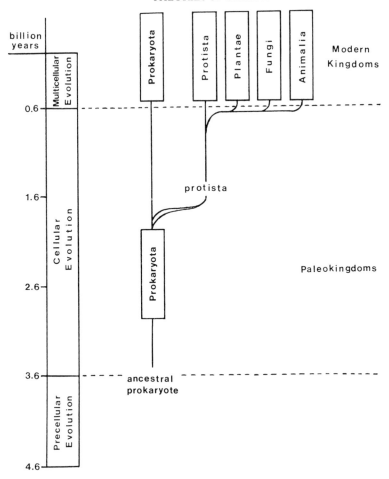

FIGURE 8:   Phylogeny of the biological kingdoms according to the Prokaryotic theory.

THE AKARYOTIC THEORY

In 1970 Roger Stanier pointed out that the microorganisms which engulfed prokaryotes by endosymbiosis could not have been prokaryotes themselves because

the impenetrability of the prokaryotic cytoplasmic membrane by any object of supramolecular dimensions effectively precludes the acquisition of endo-symbiosis.

He noticed that only the cytoplasmic membrane of the eukaryotes allows processes like phagocytosis and pinocytosis, and emphasized that in all cases of endosymbiosis the host is invariably an eukaryote:

a stable endosymbiosis in which the host is a prokaryote has never been observed.

One could object that in precambrian times some prokaryotes could have had membranes of the eukaryotic type, but bio-chemistry does not allow us to play at will with subcellular struc-tures. If there was a creature that had an eukaryotic membrane, such a creature would also have needed an eukaryotic cytoplasm, and Stanier proposed just that: the ancestors of the eukaryotes were the direct phylogenetic ancestors of the eukaryotic cyto-plasm. They were cells that did not contain nuclei or cytoplasmic organelles, and from a structural point of view they looked very much like prokaryotes: smooth pieces of cytoplasm surrounded by a membrane.

The term prokaryote however has become synonymous with bacteria, and has therefore not only a structural but also a bio-chemical and a phylogenetic meaning. According to Stanier, the ancestors of the eukaryotes were non-nucleated cells like pro-karyotes, but biochemically and phylogenetically they had nothing to do with bacteria and cannot therefore be regarded as prokaryotes. Since we obviously need a new name, I have called them *akaryotes* (without nuclei).

All models which assume that eukaryotes did not evolve from prokaryotes but that both types originated independently from ancestral anucleated cells will be considered here as different versions of 'the Akaryotic theory' of cellular evolution. The theory of Stanier was the first of these models, and can be sum-marized by saying that before the origin of the protista there were two distinct paleokingdoms on earth, prokaryotes and akaryotes, as represented in Figure 9.

Stanier accepted the idea of the common ancestor, and con-cluded that the descendants of the first cells split into two distinct lines of descent very early in the history of life. This means that

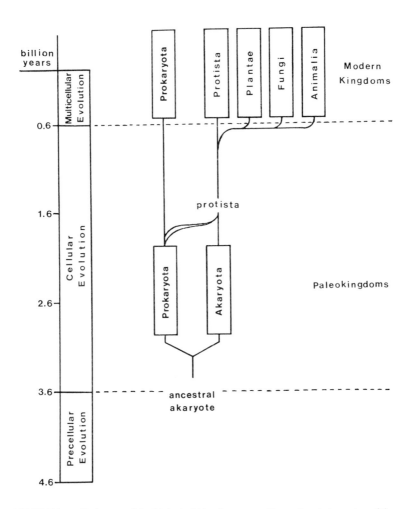

FIGURE 9:   Phylogeny of the biological kingdoms according to Stanier's version of the Akaryotic theory.

biochemically and phylogenetically the common ancestor was not a prokaryote. More than a century after Haeckel, this was the first consistent scientific alternative to the idea that life started with bacteria. It was also a major consequence of the Symbiotic theory: Stanier had to reject the Prokaryotic theory precisely because he had accepted that mitochondria and chloroplasts were acquired by endosymbiosis.

THREE PRIMARY KINGDOMS

In 1977 Carl Woese discovered that the ribosomal-RNAs are probably the oldest biological molecules on earth. The idea that we can 'date' molecules and build genealogical trees for them may appear a bit fantastic, but the underlying concept is simple and the techniques to test it are rigorous.

Take the protein cytochrome-c, for example, and look at the precise sequence of its aminoacids in different species. The sequences are very similar but not identical, and their differences reveal a pattern. Between man and monkeys, for example, the differences are small, and so are those between ducks and pigeons, but between monkeys and ducks they are significantly greater. We can measure them, obtain numbers, and draw a diagram where the gaps between species are proportional to these numbers. What comes out looks precisely like a genealogical tree.

In the case of the twenty species where cytochrome-c was actually sequenced, we have a good number of fossils whose ages are known from radioactive dating and whose relationships have been established by comparative anatomy. The paleontological tree of these species therefore is known, and if we compare it with the molecular tree we find that the correspondence is remarkably good.

The same procedure gives similar results with other proteins, and we conclude therefore that the study of sequence similarities or 'homologies' (a field pioneered by Emile Zuckerkandl and Linus Pauling in 1965), can indeed be used to build the genealogical trees of the molecules. Some may say that this conclusion is obvious (species which diverged earlier simply accumulated a greater number of variations in their proteins), but the experimental proof that this is indeed the case cannot do any harm, and is a safer starting point.

With this in mind, Carl Woese studied the sequence of nucleotides in the ribosomal-RNAs of many species, and obtained various important conclusions. The first is that the ribosomal-RNAs are the most conservative of all molecules examined. Their sequences are extremely uniform, which means that they changed very little during the history of life, or at least that they changed less than any other molecule. This is why they are said to be the oldest biological molecules on earth.

The extreme uniformity of the ribosomal-RNAs is documented by the fact that they form only three major groups. One contains the ribosomal-RNAs of all the eukaryotes, while the other two are found among the prokaryotes. This came as a surprise, because morphologically the prokaryotes are much more uniform than the eukaryotes, but the message from biochemistry is uncompromising: if the molecular sequences have a meaning (and there is ample evidence that they do), the prokaryotes do not form a homogeneous group and did not evolve in one line of descent.

This was the second major conclusion obtained by Woese: there are two distinct prokaryotic kingdoms, and he called them 'archaebacteria' and 'eubacteria'.

A third major conclusion was obtained by comparing the ribosomal-RNAs of all three groups — archaebacteria, eubacteria and eukaryotes. Each of these groups turned out to be equally different from the other two, and we have seen that equal amounts of sequence-differences mean equal amounts of phylogenetic differences. The inevitable conclusion is that the ancestors of the eukaryotes (Woese calls them 'urkaryotes') did not evolve from prokaryotes and did not appear later than prokaryotes. Urkaryotes, archaebacteria and eubacteria originated therefore independently from a common ancestor very early in the history of life, and have evolved along separate lines of descent ever since. They were the three primary kingdoms of cellular evolution.

Woese regarded the urkaryotes as the non-nucleated and non-bacterial ancestors of the eukaryotic cytoplasm, in other words as akaryotes, very much as Stanier did. The common ancestor however had to split into three paleokingdoms, not just two, and Woese's model is therefore a new version of the Akaryotic theory which is represented in Figure 10. Apart from this, Woese's evidence of the ribosomal-RNAs and Stanier's evidence of the cytoplasmic membranes converge to the same conclusion: the prokaryotes were not the direct phylogenetic ancestors of the

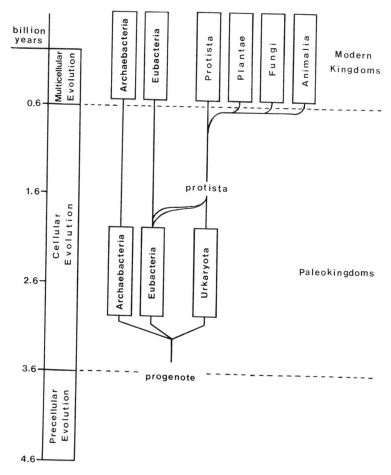

FIGURE 10:   Phylogeny of the biological kingdoms according to Woese's version of the Akaryotic theory.

eukaryotes, and bacteria were not the first cells which appeared on earth.

THE EUKARYOTIC THEORY

The evidence for symbiosis is strong, but the very evidence which proves that mitochondria and chloroplasts were acquired by symbiosis (the discovery of bacterial molecules in them), proves also that the nucleus was not acquired by symbiosis (there are no bacterial structures in the nucleus, and typical nuclear systems like nucleoli, nucleosomes, centrioli and mitotic spindles do not exist in prokaryotes). Endosymbiosis therefore explains the acquisition of some organelles, but does not account for the very thing which defines the eukaryotes.

In this situation, some have proposed a radical alternative. There is no need to explain the origin of the nucleus if we assume that a nucleus has been there all the time, i.e. that the very first cells were eukaryotes. This is 'the Eukaryotic theory' of cellular evolution, a diagram of which is shown in Figure 11. According to this theory, the prokaryotes originated *inside* the eukaryotes as organelles, were later expelled from their hosts (the reverse of symbiosis), and finally became free-living creatures. They were first dependent and then independent, not the other way round. The trouble with this theory is that eukaryotes need oxygen, and for the first two thousand million years of the earth's history there was no free oxygen on our planet. It is also difficult to accept that the most complex type of cell was the one which appeared first.

We have therefore two mechanisms (Direct Filiation and Symbiosis), three theories (Prokaryotic, Akaryotic and Eukaryotic) and four phylogenetic schemes (Figures 8, 9, 10 and 11), but no satisfactory solution for the origin of the nucleus. We can legitimately say that the origin of the cell and the origin of the nucleus are the first two entries on the list of precambrian mysteries.

THE ANCESTRAL SPLIT

According to the Prokaryotic theory the acquisition of cytoplasmic organelles and of the nucleus marked the development of the eukaryotic cell. The discontinuity between prokaryotes and eukaryotes, therefore, became an historical reality *after* this process was completed. According to the Akaryotic theory, on the contrary, the split between the ancestors of prokaryotes and eukaryotes took place long *before* the acquisition of any organelle, because that split was the precondition for the evolution of the two different types of cell that take part in symbiosis. Notice that

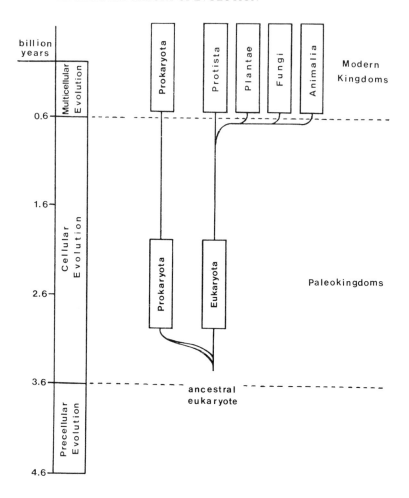

FIGURE 11:   Phylogeny of the biological kingdoms according to the Eukaryotic theory.

Stanier and Woese could not say what the ancestral split was about: they concluded that it must have taken place because they accepted the idea of the common ancestor, but that is as far as anyone can go with the Akaryotic theory.

In these circumstances, it is not surprising that there is no general agreement about the nature and the very existence of a dichotomy between prokaryotes and eukaryotes. More than a century of microbiology, cytology and biochemistry have accumulated so many facts in favour of the dichotomy that this has been described as "the greatest evolutionary discontinuity in the biological world" (Stanier, Doudoroff and Adelberg), and "the clearest, most effectively discontinuous separation of levels of organization in the living world" (Whittaker). And yet, this greatest of all biological differences is an unexplained 'accident' in modern biology.

Some have gone back to the theory of Direct Filiation, to the idea that the dichotomy is really an illusion and there is only one type of cell in Nature, with no symbiosis and no ancestral split, as if the differences documented by the molecular phylogenies had little or no meaning.

Another alternative, proposed by Woese, is the idea that there are neither one nor two types of cell in Nature but three, and therefore that the dichotomy does not make sense any longer. It is one of those concepts, like the dichotomy between animals and plants, that we should get rid of. Here, however, one can make a comparison with a similar case. The discovery that there are genetic combinations which differ from those that are typical of male and female did not mean that sex is no longer based on a natural dichotomy. In a similar way, Woese's discovery that there are three lines of descent is not incompatible with the existence of a natural split between prokaryotes and eukaryotes. Archaebacteria and eubacteria are both non-nucleated cells, and a dichotomy between cells which have and have-not a nucleus, remains. At any rate, an ancestral split into three types of cells instead of two is even more of a split, and the problem therefore does not go away.

So far there has been no theory which explains or explains away what appears to be a natural and profound discontinuity between prokaryotes and eukaryotes. The origin of this discontinuity, therefore, will become the third entry in our list of unsolved mysteries in cellular evolution.

WOESE'S DOGMA

A few years ago the evolution of the ribosome was barely mentioned in the literature, and credit for showing that this is one of the most important problems in evolution goes to Carl Woese. I would not like to underestimate the contribution of other authors, but Woese's study of the ribosomal-RNAs, in 1977, was truly a turning point: it gave us the biochemical evidence that high-molecular-weight ribosomal-RNAs, *and therefore high-molecular-weight ribosomes*, appeared very early in the history of life.

With that solid message from biochemistry, the evolution of the ribosome could no longer be ignored; it became an historical reality, and had to be accounted for. Furthermore, Woese showed not only that the event took place early in the history of life, but explained also what 'early' means in this case, and why the event was so important.

He pointed out that heavy ribosomes are essential to ensure a one-to-one correspondence between genotype and phenotype, because low-molecular-weight machines are inevitably affected by thermal noise and would make too many mistakes. A gene would not be translated into a specific protein, and the cell therefore would have a low degree of biological specificity, if any at all.

There exists a direct correlation between the size of an automaton — as measured roughly by number of components — and the accuracy of its function. ... To function accurately the ribosomes must be nearly immune to thermal noise, and so must be properly large (Woese, 1980).

Heavy ribosomes therefore had to appear early in the history of life, and early means *before the origin of biological specificity*. Every time we look at a fossilized precambrian cell and recognize fine morphological details in its structure, we can be sure that that cell had contained high-molecular-weight ribosomes. From what we know of the fossil record we conclude that the middle Precambrian was already too late: biological specificity must have appeared earlier than that, either in the first third or in the first quarter of the Precambrian.

Another contribution made by Carl Woese came from a revolutionary model that he proposed in 1970. Up until that time, it was thought that the active components of the ribosomes are their proteins, but Woese showed that the ribosomal-RNA is not an inert scaffolding and has a direct functional role. According to his 'ratchet model', the very essence of the translation machine lies in the interaction between two allosteric pieces of RNA. If we imagine taking those two pieces out of the machine and stabilizing them with small peptides, we can visualize what a very small ribosome would look like, and we can start making models of low-molecular-weight ribosomes.

In short, Woese made it possible to think of very small ribosomes in realistic terms, noticed that the evolution from light to heavy ribosomes is a precondition for biological specificity, and gave us the biochemical evidence that heavy ribosomes appeared early in evolution. These three ideas can be put together in what I call 'Woese's dogma': *the evolution from low to high molecular weight ribosomes took place very early in the history of life, before the origin of biological specificity.*

This is, in my opinion, a conclusion of the greatest importance: if a primitive cell had biological specificity, we simply cannot say that that cell had genes and proteins without also saying that it had heavy ribosomes which translated genes into proteins with accuracy.

PROGENOTES AND HYPERCYCLES

Having established that the evolution of the ribosomes from low to high molecular weights took place at the very dawn of life, Woese proposed a theory which accounts for that event with three hypotheses (the Progenote theory).

a) The increase of the ribosome molecular weights was favoured by natural selection because it increases the accuracy of translation. Ribosomes became heavy in order to become precise.

b) Since the increase of the ribosome molecular weights was favoured because it creates a one-to-one correspondence between genotype and phenotype, such increase had to take place *within* a genotype-phenotype entity, i.e. within a cell. The very first cells which appeared on earth, therefore, had low-molecular-weight ribosomes. As a consequence they had little specificity: any gene was translated not into identical proteins but into a class of 'statistical proteins', all more or less different from each other. Woese called these primitive cells *progenotes.*

c) Progenotic evolution ended, and eugenotic (or specific) evolution began, when the ribosomes reached molecular weights of about 1 or 2 million, values which correspond to the types of ribosomal-RNAs which have survived ever since. This means that the evolution from high to very-high molecular weights (from 2 to 4 million or more), took place later, and took place also within the cell.

I would like to distinguish sharply between Woese's dogma and the Progenote theory, because I accept the first but not the second. In my opinion, the Progenote theory is not a satisfactory solution of the ribosome problem for the following reasons.

A) Prokaryotes have 70S ribosomes whose molecular weights are just over 2 million, while the 80S ribosomes of the eukaryotes often exceed 4 million. The increase from 2 to 4 million, however, cannot be explained by an increase in accuracy because prokaryotic and eukaryotic ribosomes translate with the same accuracy. The idea that natural selection favoured the increase in order to improve accuracy cannot be true in this case.

Could it be true in the other case, for the increase from a few thousand daltons to 1 or 2 million? We cannot really accept this, because Nature does not give rewards for future advantages. Consider an equivalent case. What was the advantage of increasing 5% of a wing to 10%? Since neither value could be used for flying, it is obvious that there had to be another reason for the development of the wing, a reason which had nothing to do with flying.

In a similar way, the advantage of an accurate translation system could be appreciated only when the molecular weight increase had gone full-way to 1 or 2 million. Nature therefore must have favoured the evolution from light to heavy ribosomes for reasons which had nothing to do with the accuracy of translation.

B) The second hypothesis of the Progenote theory is that the increase from low to high molecular weights of the ribosomes (up to 1 or 2 million), took place within the cell. The very first cells had small ribosomes and therefore could only produce statistical proteins.

My objection is that statistical proteins can be specified by slightly different genes, and the progenotes therefore would have not only statistical proteins but also statistical genes. The two inevitably go together and, what is worse, they tend to widen each other's range. Any slightly different statistical protein would widen the range of the statistical genes, and any slightly different statistical gene would widen the range of the statistical proteins. This means that a low degree of biological specificity tends to become lower, not higher, and eventually the cell would face total chaos. The progenotes would have become extinct fairly quickly.

C) The third hypothesis of the Progenote theory is that the evolution of the ribosome from high to very-high molecular weights (from 2 to 4 million or more) also took place within the cell. Let us assume that this is true, that the first cells did have ribosomes of 1 or 2 million which translated with accuracy. Any mutation which caused a further increase in their weight would not have increased their translation accuracy, but would surely have increased the metabolic burden on the cell, and eventually any substantially increased burden would have been selected against. Even the third hypothesis therefore is unsatisfactory.

In short, if the first cells were the little bags full of low-molecular-weight chemicals that Woese calls progenotes, we have good reasons for concluding that they did *not* acquire heavy ribosomes and did *not* evolve towards biological specificity. To say that they must have done so because life did appear is not good enough. Unless we have a convincing mechanism, the hypothesis that life started in little bags full of chemicals is no starting point at all.

Another model which is based on this hypothesis is 'the Hypercycle theory' of Eigen and Schuster (1977). The hypercycle is "a cyclically closed hierarchy of single reaction cycles ... whose components had to be surrounded by a membrane. Such a complex reaction network may already be called 'live' and could stand for a primitive precursor of a single cell". Furthermore, the crucial chemicals of Eigen and Schuster are small macromolecules like transfer-RNA and activating enzymes.

The Hypercycle theory therefore implies that life evolved in little bags — or in equivalent systems — like the Progenote theory, even if the details are quite different. Eigen and Schuster, for example, do not even describe the evolution of high-molecular-weight ribosomes, but that is the real obstacle of all the theories of this class and ignoring it is no good. To say that vesicles containing organic chemicals appeared in the primitive solutions is one thing. To show how these little chemicals evolved into huge translation machines by means of natural processes, is a totally different proposition.

The evolution of the ribosome, in conclusion, is not accounted for by the Progenote theory or by the Hypercycle theory or by any other theory of the same class. I will regard it therefore as entry number four in our list of unsolved problems of pre-cambrian evolution.

THE DIVORCE BETWEEN ORIGINS AND EVOLUTION

Many biologists would say that the origin of the cell is the greatest of all biological problems, and yet there is still a wide-spread belief in a fundamental divorce between the Origin and the Evolution of life. Darwin himself said that "Evolution has nothing to do with the Origin of Life", and today this conclusion is much more popular than people care to admit. The divorce is based on this reasoning: the origin of the cell was such a *turning point* that what happened before did not matter any longer. As a result, we can reconstruct and understand the evolution of the cell whatever was the cause of its origin.

If this is true, clearly there would be a divorce 'in principle' between Origins and Evolution, and I think that many biologists do believe in the 'turning point' idea. This is why the Origin of Life is something which remains at the periphery of Biology, something that in principle is terribly important but that in practice is not necessary to understand Evolution. This also explains why the theories of Sudden Life are popular. The first cells could have come from Space, could have been created by God or could have appeared in any other way; it would make no difference to what happened afterwards.

I began to have doubts about the 'turning point' idea when I realized that there may be a connection between the origin of the cell, the origin of the nucleus, the split between prokaryotes and eukaryotes, and the evolution of the ribosome. Naturally, a link between these four precambrian problems would mean that there is a link between what happened before and after the origin of the cell. It would mean that there is no divorce between Precellular and Cellular Evolution, that we cannot really hope to understand the evolution of the cell without a consistent theory about its origin.

The theory that I published in 1981 in the Journal of Theoretical Biology and which is described in the next chapter, is what came out of this line of research.

CHAPTER 7

The Ribotype Theory of the Origins

RIBOSOIDS AND RIBOTYPE

In 1981 I introduced the term *ribosoids* to indicate all molecular systems which contain the sugar *ribose*. The ribosoids therefore can be either pure RNAs or compounds of RNA and other molecules, for example ribonucleoproteins. They can have low or high molecular weights, and their association with aminoacids or proteins can be either temporary or permanent. ATP, transfer-RNA and ribosomes, for example, are all ribosoids. Furthermore, I gave the name of *ribotype* to the system formed by all the ribosoids of a cell, and proposed 'the Ribotype theory' on the basis of two concepts: the cell is not a duality of genotype and phenotype but a trinity genotype-ribotype-phenotype, and life on earth originated with the ancestors of today's ribotypes.

The Ribotype theory of the Origins will be developed gradually in this chapter, but I would like to clarify two points at the very beginning. The idea that RNA preceded DNA in evolution is not new and should not be confused with the ribotype concept. It is a version of the Genotype theory, because it regards the RNAs as the first replicating molecules which simply anticipated the role of DNA. In the Ribotype theory, instead, the critical role is attributed to the ancestors of the ribosomes, not to the ancestors of the genes.

The second point is that the trinity genotype-ribotype-phenotype is not a repetition of the Central Dogma DNA → RNA → Proteins, because the RNA of the Dogma is messenger-RNA, not ribosomal-RNA. The relationship between genotype, ribotype and phenotype cannot coincide therefore with the flow of biological information of the Central Dogma, and it will be shown later that it has indeed a more general nature. With these notes and the above terminology, let us now go back to precellular evolution and discuss a new approach to it.

PROTORIBOSOMES

Like many other models on precellular evolution, the ribotype theory starts from the hypothesis that various types of organic molecules were formed spontaneously in the solutions of the primitive earth. The next step is to notice that if peptides and nucleotides were present, ribosoids were also present, because ribonucleic acids and aminoacids interact so easily that in a random mixture it is more likely to find them together than apart. In simulation experiments various types of ribosoids have been obtained, and the spontaneous formation of analogous compounds in primitive solutions is therefore a plausible assumption. Furthermore, Thomas Cech has recently shown that some ribosomal-RNA molecules can perform biochemical reactions upon themselves, behaving as self-catalysts, and it is not unreasonable therefore to assume that primitive ribosoids had, in a crude way, similar active characteristics.

Let us now discuss what very primitive ribosomes, or proto-ribosomes, could have been like. I borrow from Woese the idea that a few pieces of RNA stabilized by small peptides can be regarded as protoribosomes, but the gap between these ultra-simplified systems and their modern descendants is enormous and deserves a very careful examination. More precisely, we should take great care in discussing which properties of modern ribosomes can be legitimately attributed to their small ancestors. As a starting point, let me quote a metaphor from Jacques Ninio.

"Returning late to my hotel, I try to open the door of my room, but the lock refuses to yield. I then noticed that I had mistaken the door. Finding the right door, I open it without difficulty and admire the fine indentations on the key, imagining that the lock bears inside it the same complicated zigzags in reverse. The next day, the chambermaid opens all the doors on the corridor with calm assurance and an almost magical key: the pass-key. It might be thought that the pass-key is even more elaborately indented than mine since it opens all the doors. But no, it is a smooth and everyday key". The reason is simple. It does not take much to open doors: a screwdriver or even a piece of plastic is enough. What is complicated is to open one door but not the others, the ability to recognize and discriminate.

Today's ribosomes are not only heavy and complex, but

require a whole battery of activating enzymes to recognize, discriminate and arrange molecules in a specific order. Let us assume however that all the protoribosomes could do was to stick aminoacids together at random, in other words that they were not translation machines but mere 'polymerizing' machines, or 'polymerizing ribosoids'. For these systems we can indeed expect much less than for their modern counterparts. For a start, we do not need high molecular weights to combat thermal noise and reduce errors because a random sequence with errors is still a random sequence; the very concept of error does not exist at this level. It is legitimate therefore to assume that protoribosomes had very low molecular weights.

Furthermore, we do not need recognition enzymes, because no recognition or discrimination was necessary. By the same token we do not need messengers, because the translation of a random sequence of nucleotides would still be a random sequence of aminoacids. And if we do not need messengers we do not need a genetic code.

In short, the enormous difference in weight and complexity between protoribosomes and modern ribosomes can be explained naturally by the difference between what is necessary to create a random and a specific sequence of aminoacids. The protoribosomes were the small pieces of plastic which are only capable of opening doors. The modern ribosomes are complicated sets of keys which open specific doors in a specific order.

My conclusion is that translation would be very difficult to justify at an early stage, but polymerization does not present difficulties. The ability to polymerize aminoacids at random is therefore the first property that I attribute to the protoribosomes. More precisely, I conclude that in a primitive population of random ribosoids some were 'polymerizing ribosoids', systems which could catalyze the formation of a peptide bond between any two aminoacids. This presents us with a problem: how likely was the spontaneous appearance of protoribosomes? Was it a very rare event or a relatively common one?

SELF-ASSEMBLY AND POLYMORPHISM

In addition to the ability to polymerize, I attribute to the protoribosomes the properties of self-assembly and polymorphism. We know that modern ribosomes can be assembled in vitro from their components, and I assume that protoribosomes were also capable of self-assembly because small ribosomes are reconstituted in vitro more easily than heavy ones, and in general because self-assembly occurs more readily in simpler molecular systems. At any rate, the hypothesis is legitimate because self-assembly is a well documented property of the ribosomes, and by attributing it to the protoribosomes I am not inventing anything extravagant or improbable. I am simply assuming true in the past what we know to be true today.

For the same reason I attribute to the protoribosomes the property of polymorphism that is so common among modern ribosomes. By polymorphism, in this case, I mean the property by which ribosomes of different species can have components that differ in number, molecular weight, physico-chemical characteristics and three-dimensional structure, and yet they are all ribosomes, they all perform the same function with the same accuracy 'as if' they were identical.

More precisely, by saying that protoribosomes were polymorphic, I mean that their RNA molecules could have different lengths and could be stabilized by different peptides, but were all capable of sticking aminoacids together. They could polymerize at random despite their structural differences, just as modern ribosomes can polymerize with specificity despite their structural differences.

Let us now examine what chance protoribosomes had to appear spontaneously in primitive solutions. The property of self-assembly means that protoribosomes of a certain complexity (a few thousand daltons, for example) could originate from small molecules in a straight-forward way. The property of polymorphism means that there were many different ways of building them, the higher the polymorphism the greater the statistical frequency of the protoribosomes, and therefore their chance of appearing.

We cannot put a number on this, but we know that there are an enormous variety of individual differences among modern ribosomes, and it is safe to assume that there was a very substantial degree of polymorphism among their ancestors. I conclude therefore that in a random population of ribosoids the percentage of polymerizing ribosoids was presumably small, but not so small as to be insignificant. This is only a qualitative conclusion, but it is nevertheless valuable: it means that the appearance of protoribosomes in primitive solutions was not an extremely improbable event, but rather a likely one.

QUASI-REPLICATION

The presence of polymerizing ribosoids in the primitive solutions ensured that many random proteins came into existence, and in this situation we do not need to rely on improbable accidents to assume that proteins of a particular class were formed. Given a large enough number of random products, any variety was bound to appear with percentages and frequencies that are determined by statistical factors only. Some proteins, for example, were bound to be of the kind that favour the condensation of random chains of nucleic acids, which means that the production of polynucleotides was also accelerated.

New polypeptides and new polynucleotides lead, among other things, to new ribosoids, and we have seen that in any population of random ribosoids a small but significant percentage is formed by polymerizing ribosoids. We have therefore a cycle here, whereby a generation of protoribosomes leads to new random proteins, new random nucleic acids, new random ribosoids, and finally to a new generation of protoribosomes that start the cycle all over again.

It should be noticed however that such a cycle has a very peculiar nature: we cannot speak of replication, not even of replication with errors, because there is no copying mechanism, no template and no transfer of information (this is very different from the 'statistical proteins' of Woese and from the 'quasi-species' of Eigen and Schuster, where the differences are due to replication with errors). In any generation, here, the components of the protoribosomes can all differ from the components of the previous generation not because errors are made, but quite simply because the system is not programmed to reproduce its original components. Despite this, a new generation of protoribosomes does come into existence, because polymorphism gives the same polymerizing capability to a vast class of ribosoids, and because the polymerizing ribosoids are capable of self-assembly.

We have here a new process that I have called *quasi-replication*, and we can recognize a quasi-replicating system by three characteristics. The first is that only a fraction of its products are reinvested in the making of a new system (enormous waste). The second is that the new systems are put together by self-assembly (no transfer of instructions). The third is that the components of any new generation are likely to be structurally different from the components of the previous generation (polymorphism).

Quasi-replication allowed the protoribosomes to perpetuate themselves, and it did not really matter if they could not produce replicas of their own components. The important point was that new protoribosomes could carry on the same function as their predecessors 'as if' they were identical.

NUCLEOSOIDS

In addition to polymerization, self-assembly and polymorphism, I attribute to the primitive ribosoids a fourth property: the ability to form irregular clusters of supramolecular dimensions. Ribonucleic acids and ribonucleoproteins can form aggregates and coacervates of various kinds, and the clustering ability of the ribosoids is yet another property for which we have ample evidence. It is only natural therefore to assume that such clusters could and did originate in the primitive solutions, particularly when these became enriched by protoribosome-driven synthesis of proteins and by enzyme-driven synthesis of nucleic acids.

The limits to the dimensions of these aggregates is anybody's guess, but the best example that we have today is represented by the nucleoli, and because of this I have called them *nucleosoids*. Nucleosoids are, therefore, clusters or coacervates of ribosoids, or of ribosoids and other molecules, and can have a variety of sizes, shapes and compositions. From this general definition it follows that the nucleosoids of the primitive solutions were enormously heterogeneous, and since they were formed by random processes there is no doubt that the great majority were inert or nonsense structures, dead-ends with no evolutionary future. The same statistical argument, however, leads us to conclude that a small percentage had more interesting features.

We can have an idea of these features by generalizing four characteristics of modern nucleoli.

1) A three-dimensional network of ribonucleoproteins is not necessarily an inert structure, but can allow physical movements and biochemical transformations, like nucleoli do.

2) A three-dimensional network of ribonucleoproteins creates microenvironments and microcompartments which trap molecules and localize their interactions.

3) Clusters of ribosoids can have all the intermediate dimensions between the angstrom and the micron. They provide a bridge between molecules and cells.

4) Nucleoli of different species are very different in size, shape and components, and yet they are all nucleoli, they all behave as if they were indentical. They have supramolecular polymorphism as the ribosoids have molecular polymorphism.

Internal movements, internal compartments, a wide range of dimensions and supramolecular polymorphism: these are characteristics of the nucleoli that we can extend to the larger class of the nucleosoids. It may be pointed out that nothing has value in evolution if it has no lasting effects, and however interesting the nucleosoids are, we cannot attribute replication properties to them. This is true however only if we are speaking of true replication, of replication based on a copying mechanism, with or without errors. Quasi-replication is a very different matter, and we should examine it carefully.

We have seen that quasi-replication is based on three properties: ability to polymerize, self-assembly and polymorphism. Two of these are already present in the nucleosoids: they can assemble themselves by clustering, and their supramolecular polymorphism ensures that they can do so in many different ways.

The ability to polymerize, instead, was not present in all of them, but we have seen that in a population of random ribosoids a small but significant percentage is made of polymerizing ribosoids. It is legitimate therefore to conclude that a small percentage of nucleosoids could and did contain polymerizing ribosoids. These were the interesting ones, because they had all three attributes of a quasi-replicating system, and were therefore capable of leaving descendants.

A nucleosoid which could produce other ribosoids would pre-ferentially make its own components, and could grow on itself reaching whatever dimensions were physically attainable. Even-tually, however, it would become unstable, break apart into smaller pieces, and in some of these the ribosoids that were responsible for the previous syntheses would go on repeating the cycle. The polymerizing nucleosoids could leave descendants by quasi-replication, and however small their initial number was, they had the potential to outlive all other nucleosoids. They had a future.

HETEROSOIDS

The ideal end of nucleosoid evolution may appear to be a situ-ation where nucleosoids produce exclusively their own ribosoids, but this is not a realistic outcome. There is an inherent random-ness in their syntheses, and a whole variety of non-ribosoidal components were bound to be produced at all stages. One of these, for example, is represented by the membranes. If the asso-ciation with a membrane was not too unstable on physico-chemical grounds, a subclass of membrane-associated nucleo-soids was bound to become established.

The most important sort of 'contamination', however, was DNA. Molecules of DNA were likely to exist in the primitive solutions, and could easily be formed from RNA by removing one oxygen from the ribose, but they did not have an important role in the first stages of evolution, because they lack the poly-morphism which is essential for quasi-replication. They are very stable, one-purpose molecules, and their properties became use-ful only at an advanced stage, when some nucleosoids started carrying instructions and learned to use templates for their syn-theses. RNA is perfectly capable of storing instructions, but here DNA does a better job, precisely because it lacks flexibility, is less reactive and more stable. At any rate, the probability that some nucleosoids could produce or trap DNA was high, and it can safely be assumed that a significant number of them became closely associated with DNA molecules.

E

We can see therefore three major developments taking place during the evolution of the nucleosoids. In some of them DNA started appearing in the inside, in others membranes started forming protective coats on the outside, and in a smaller subclass of nucleosoids *both* developments took place. In these subclasses the relationship between ribosoids and non-ribosoids became increasingly close and I summarize this process by saying that some nucleosoids became heterogeneous-nucleosoids, or *heterosoids*.

The first heterosoids, however, were not yet the first cells. Their quasi-replication mechanisms were still dividing them into unequal parts, and a new phase of evolution had to take place before some could produce equal descendants. When this phase was also completed, quasi-replication became true replication. Precellular Evolution came to an end, and the first cells appeared on Earth.

THE EVOLUTION OF THE RIBOSOMES

Let us see now if we can be a little more specific about some pre-cambrian developments, in particular about the evolution of the ribosomes from low to high molecular weights. We have seen that the appearance of small polymerizing ribosoids, or proto-ribosomes, was not a highly improbable event in the primitive solutions. Let us start from these very small machines and ask the following question: was there a limit to the length of the chains that they were producing? As far as we know, they could go on sticking subunits together indefinitely, and could produce polypeptides and polynucleotides of low, medium, high and very high molecular weights. The production of a short chain is not inherently different from that of a long one, and we cannot say therefore that heavy molecules had necessarily to appear *very late* in precellular evolution. Nucleosoids could well have appeared shortly after the polymerizing ribosoids, and even if they were inert clusters, at the beginning, they had nevertheless a useful role to play.

Compare for example the self-assembly of a ribosoid in an open solution and inside the microenvironments of a supra-molecular cluster: it is natural to conclude that the process was much quicker, on average, within the clusters, however inert they were. Furthermore, a polymerizing machine was more likely to go on undisturbed and produce heavy molecules within a cluster than in a free-floating state. After their appearance, therefore, the supramolecular clusters were likely to be the places where most of the new heavy molecules were actually produced, and where a variety of trial-and-error experiments were taking place.

Let us assume now, merely as an example, that the polymerizing machines of two nucleosoids produced chains that were, on average, 50 and 500 subunits long. The first nucleosoid would have smaller ribosoids than the second and therefore smaller protoribosomes, on average, but self-assembly was a common property and the appearance of light protoribosomes in the first nucleosoid was as natural as the appearance of heavy protoribosomes in the second.

Nucleosoids with protoribosomes of different molecular weights, therefore, could well have existed side by side, but this does not mean that they were equivalent. Their abilities to polymerize were presumably similar, but the average dimensions of their clusters were not, and this did make a difference. Nucleosoids formed by heavier particles were statistically bigger, and clusters of larger dimensions had various advantages: they could trap more organic molecules and perform more trial-and-error experiments per unit of geological time. In short, they could evolve faster.

One may say that the connection between the molecular weights of the protoribosomes and the dimensions of their supramolecular clusters is not obvious, but we have today at least one case which illustrates it: the ribonucleoproteins of the 80S class give origin to nucleoli when they are ribosome precursors, and mature into synthesizing ribosomes when their biogenesis is completed. In this case, the production of ribosomes goes hand in hand with the production of nucleolar clusters, and something very similar could well have been true for protoribosomes and nucleosoid clusters.

This allows us to reach a new conclusion about precambrian ribosomes: the evolution from low to high molecular weights did not take place within the cell but at the precellular level, and was not favoured by accuracy in protein synthesis, because small and large protoribosomes could polymerize equally well, as today small and large ribosomes can translate equally well. It was favoured instead by the characteristics of supramolecular clustering, because bigger clusters can trap more molecules, can perform more reactions simultaneously, and can leave more descendants by quasi-replication in the same period. We have in this way a consistent solution to one of the four precambrian mysteries which were listed in the previous chapter.

MICROKARYOTES

So far I have attributed to the ancestral ribosoids four properties (ability to polymerize, self-assembly, polymorphism and supramolecular clustering), but there is at least one more characteristic that should be taken into account. Molecular phylogeny has shown us that the class-difference between prokaryotic (70S) and eukaryotic (80S) ribonucleoproteins has been strongly conserved during evolution, and something very similar therefore must have existed at the dawn of life. This primitive 70S-80S class-difference is the fifth and last property that I attribute to the ancestral ribosoids, and from it we can learn quite a few things about early life.

Modern prokaryotes do not have nucleoli, and if a general 70S-80S difference existed in the past we conclude that primitive prokaryotic ribonucleoproteins were as unable to form proper supramolecular clusters as their modern descendants are. This means that true nucleosoids could only have come from the ancestors of a primitive 80S type, or a primitive higher type that we may call 90S. Only these heavy ribonucleoproteins had the ability to form supramolecular systems capable of reaching the dimensions of a micron, and to create microcompartments that could trap a wide variety of organic molecules.

In this way the Ribotype theory concludes that the first cells were not bacteria, because they did not have primitive bacterial ribosomes and did not evolve from ancestors which had bacterial ribonucleoproteins. They evolved from nucleosoids, and contained therefore a primitive nucleolus which was at the same time their supporting structural framework and their centre of activity. For this reason I have given the first cells the name of *microkaryotes* (containing a diminutive nucleus or, more precisely, a primitive nucleolus). This sharply differentiates the ribotype model from all versions of the prokaryotic and akaryotic theories, where the first cells are described as optically transparent little vesicles, i.e. as primitive bacteria.

Notice that a nucleolus surrounded by a membrane can measure less than a micron, and can easily be smaller than the smallest bacteria. This means that the microkaryotes had no organelles such as mitochondria and chloroplasts, and did not divide by mitosis; as a consequence they did not require oxygen and were fully anaerobic. The microkaryotic model, in other words, has nothing to do with the eukaryotic theory of the ancestral microorganisms.

The image of the first cells as bundles of ribonucleoproteins surrounded by membranes is a new model, but we should not regard it as far-fetched. As we have seen, Carl Woese concluded that primitive ribosomes of the 80S class appeared very early, and my hypothesis takes this conclusion a step further: I assume that the cells which had primitive 80S ribosomes also had primitive nucleoli, simply because nucleoli are *formed* by 80S ribosome precursors. We realize in this way that cells with primitive nucleoli must have appeared on earth extremely early, and could well have been the creatures which started life on our planet.

PROKARYOTES AND MICROEUKARYOTES

Although the microkaryotes could easily have been smaller than bacteria, their primitive nucleoli must nevertheless be regarded as bulky and inefficient structures, because they were the result of quasi-replication processes that are inherently wasteful. When a true replication strategy replaced the old mechanism, therefore, readjustments were only too natural and the ancestral ribotypes could have been streamlined in various ways. How many?

The most direct way of streamlining a system is to get rid of all unnecessary components, and I assume that Nature tried all sorts of experiments in this direction. Ribosome biogenesis could be simplified and various genes for ribosomal proteins could be eliminated altogether, while others could be reduced in size to code for smaller proteins and therefore smaller ribosomes. When the molecular weights become too low, however, the ribosomes become too sensitive to thermal noise, make too many mistakes and the cell loses biological specificity. The 'devolution' towards low molecular weights, therefore, had a natural limit in the need to maintain a high level of accuracy in translation and, judging from the results, the prokaryotes emerged as the best compromise.

Of all the experiments that Nature could try in this direction, it would be unreasonable to assume that only one could succeed. Archaebacteria and eubacteria are two distinct classes of prokaryotes, and I interpret their phylogenetic diversity as the result of two different routes that Nature followed in the dismantling of the ancestral ribotype. The fact that both classes have 70S ribosomes reinforces the conclusion that these are the smallest translation machines which ensure biological specificity.

The descent towards 70S ribotypes, however, was only one of Nature's options. The overall efficiency of the microkaryotes could also be improved without a massive dismantling of their ribotype; in fact, small readjustments are more likely to pay off in the short term, and I conclude that they too were made. We see therefore that a solution is taking shape. The first cells were carrying a cumbersome and inefficient ribonucleoprotein system, and it is reasonable to assume that at least two types of streamlining processes were under way. This is a little vague, but we can go further.

From what we know, the most natural way of streamlining a biological system was by using primitive enzymes, and these could have been 'DNA-cutting enzymes' as well as 'RNA-cutting enzymes'. This gives us two classes of streamlining processes which lead naturally to two classes of cells. The cutting enzymes of today are the result of a long chain of modifications, and we can recognize in them only remnants of their distant ancestors, but these general characteristics are nevertheless useful.

We know that prokaryotes have 'DNA-cutting enzymes' (the restriction enzymes of genetic engineering) which give some degree of protection against viruses because they can recognize foreign DNA and destroy it. It is unlikely however that the restriction enzymes were developed to combat viruses because as defence mechanisms they are very inefficient. Many viruses can fool them easily, as biologists do when they insert pieces of foreign genes in bacteria.

My interpretation is that the restriction enzymes are the modified descendants of primitive DNA-cutting enzymes that were used to streamline the genetic apparatus of some microkaryotes. Having eliminated by trial-and-error experiments the non-essential pieces of DNA, at least some cutting enzymes could still be used to attack foreign DNA. Perhaps their main purpose was not to combat viruses, but to keep the total amount of DNA to a minimum, so that nothing in the cell would be wasted and maximum efficiency could be maintained.

The ribotype theory explains therefore the origin of the bacteria in the following way: the first prokaryotes were created from some microkaryotes by a streamlining process that was primarily based on primitive DNA-cutting enzymes. Let us now look at the microkaryotes that did *not* have these enzymes. They could not get rid of their genes, including the ribosomal genes, and so the cumbersome apparatus of their primitive nucleoli was not dismantled. In these circumstances it is likely that most of their descendants became extinct simply because they were too inefficient, but some could survive. If pieces of DNA could not be eliminated, some cells could still manipulate their genes by using enzymes that make DNA from RNA by reverse transcription. This would increase the number of genes instead of decreasing them, but then the cells could use primitive RNA-cutting enzymes (the ancestors of the modern splicing enzymes) to eliminate nonsense bits before translation.

This mechanism may look clumsy and complicated compared with a direct intervention on the genes, but it had two valuable characteristics: the ability of the cell to carry a large number of genes was increased rather than decreased, and RNA-shuffling could be exploited to create biological diversity. By using some RNA-enzymes for reverse transcription and some for RNA-splicing, a cell could exploit a powerful new mechanism for assembling novel proteins and explore a whole new range of possibilities.

The hypothesis that some microkaryotes developed both types of RNA-enzymes may at first seem unlikely, but in fact we already have some experimental evidence for it. Reverse transcription and RNA-splicing have been found almost exclusively in eukaryotic systems, and it is not unreasonable therefore to assume that they were developed together in the ancestors of the eukaryotes. Other recent discoveries have shown that RNA-splicing and split-genes are phylogenetically very old, and the hypothesis that they appeared early is therefore justified.

According to the ribotype theory, in conclusion, the streamlining of some microkaryotes by DNA-enzymes led to the ancestors of the prokaryotes (archaebacteria and eubacteria), while the manipulation of others by RNA-enzymes led to the ancestors of the eukaryotes, cells that I called *microeukaryotes*. This tells us that the ancestral split was essentially a split in the ancestral ribotype, and gives us a natural interpretation for the origin of two main types of cells from one.

THE ORIGIN OF THE NUCLEUS

Prokaryotes and eukaryotes, as we have seen, are separated by a 'structural' discontinuity (the presence or the absence of a nucleus) and by a 'biochemical' gap (some molecules exist in one type of cell only). There is however a third difference between them, a 'physiological' discontinuity that represents, in my opinion, the best argument in favour of a natural dichotomy. This is the existence of two distinct relationships between transcription and translation.

The presence of a nuclear membrane means that transcription (in the nucleus) is physically separated from translation (in the cytoplasm), while the absence of such a membrane means that the two processes can be linked, in the sense that messengers can be translated at one end while their transcription is still going on at the other. As a result of this, prokaryotes can establish and exploit a physical link between transcription and translation, while eukaryotes cannot, and the presence or the absence of such a link has various important implications.

One, for example, concerns the control of protein synthesis. A strict link is compatible with short-lived messengers, while without it the cell has to use long-lived messengers, and the whole strategy for the regulation of protein synthesis is affected. A second consequence is that a strict link requires an *open* arrangement of the genome, in the sense that genes must be accessible to the ribosomes, and this puts a limit on the total number of genes that the cell can carry. A prokaryotic cell would have to be enormous in order to carry the number of genes of a typical eukaryote in a ribosome-accessible configuration.

In eukaryotes, on the contrary, the genes can be packed much more tightly because translation takes place away from them, and they are no longer required to have direct access to the ribosomes. As a result, eukaryotes can carry a higher number of genes per unity of cell volume. We can reach in this way the following generalization: prokaryotes are cells where transcription can be physically linked to translation (they can take advantage of this or not, but the potential is there); eukaryotes are cells where transcription is physically separated from translation.

This analysis however makes it all the more clear that any theory on the origin of the nucleus has to face an apparent paradox: a nuclear membrane would choke the cell to death if ribonucleoproteins could not go through it. The ability of the ribosomes to cross such a membrane, in other words, had to exist *before* the nuclear membrane came into existence. Furthermore, the cell had to have long-lived messengers in order to be able to separate transcription from translation.

These are necessary preconditions for the origin of the nucleus, but if we assume that the precursors of the eukaryotes were bacteria — either prokaryotes or akaryotes — we simply do not know how to account for them. Within the framework of the ribotype theory, there is a very natural interpretation.

A nucleolus forms a supramolecular barrier around the ribosomal genes which prevents them from being directly accessible to the ribosomes. At least for these genes, therefore, transcription was physically separated from translation. Furthermore, in the nucleolus the ribosome precursors are physically transported from one point in space to another across a supramolecular barrier.

The microkaryotes and the microeukaryotes, therefore, already had the essential physiological prerequisites: transcription was separated from translation, and their ribonucleoproteins had the ability to cross supramolecular barriers. The nuclear membrane is just that — a supramolecular barrier — and the nucleus therefore could grow around the nucleolus as its natural extension.

This is the ribotype solution. It is not a detailed description, but with it the gradual evolution of the nucleus becomes a process that we can visualize and understand naturally, without the paradoxes that confront us in other frameworks. I conclude that the ancestors of the eukaryotes were cells which had primitive nucleoli, and I feel that we can safely abandon all theories where the origin of the nucleus is totally incomprehensible.

THE DEVELOPMENT OF THE EUKARYOTES

With the bacterial organization, Nature had created a masterpiece of cellular engineering. The prokaryotes were a leaner and fitter type of cell, where nonsense instructions and other molecular waste had been eliminated or were in the process of being eliminated. By contrast, the microeukaryotes had to invest at least twice as much energy and materials in ribonucleoproteins, to obtain ribosomes which had the same accuracy. Clearly, in a competition with the prokaryotes for natural resources they would have been metabolically handicapped.

We can allow a certain amount of time for such a confrontation, because it is unlikely that the first cells expanded rapidly over the planet. The streamlining of the prokaryotes took time, their differentiation into various species also took time, and when some of them started producing oxygen, all the cells had to adapt slowly to a slowly changing environment. We know however that the prokaryotes eventually managed to colonize the earth, expanded virtually everywhere and differentiated into countless types. It would be unrealistic to assume that the oceans could sustain an unlimited growth of all kinds of microorganisms, and a confrontation between prokaryotes and the precursors of the eukaryotes was virtually inevitable.

In 1970, Roger Stanier reached the same conclusion with a different argument. He pointed out that without chloroplasts and mitochondria, the ancestors of the eukaryotes could only exploit glycolysis — the most primitive form of fermentation — for their energy requirements. By contrast, the prokaryotes had at their disposal a wide variety of more efficient metabolic mechanisms which gave them decisive advantages in all environments. For at least two different reasons, therefore, the 'emerging eukaryotes' (as Stanier called them) had all the odds against them, and yet they managed to survive. How? Stanier proposed a solution that I find extremely convincing, particularly because it gives an answer to various other problems as well.

Since the eukaryotic cytoplasmic membrane has the potential to engulf supramolecular objects, Stanier suggested that the emerging eukaryotes used that potential and developed processes like pinocytosis and phagocytosis to obtain nutrients for glycolysis from other cells. Being unable to compete metabolically with the prokaryotes, in other words, they started eating them. They became predators, and in this way solved once and for all their food and energy problems.

The model explains a variety of eukaryotic characteristics which are apparently unrelated. It accounts, for example, for a substantial increase in cell size, and for the development of structures like the microtubular system and the Golgi apparatus as a means for promoting the active locomotion of the cell (cilia, pseudopodia) and the ability to hunt, capture, devour and digest the prey. At the same time the microtubular system, once developed, provided the means for the evolution of mitosis.

Another beauty of the Stanier model is that it explains symbiosis as a natural complement to the predation mechanism. The ability to phagocyte other cells could have been dissociated from the ability or the necessity to digest them, in which case the engulfed microorganisms would have survived within their hosts. Mitochondria, chloroplasts, microtubuli, cilia, pseudopodia, Golgi and mitotic apparatus, are in this way all accounted for.

Finally, if the prokaryotes were food the emerging eukaryotes could not wipe *them* out without starving themselves to death. With a relationship of prey-and-predator the model explains in the most natural and convincing way why both types of cells survived and flourished.

There is only one point where I depart from Stanier: I believe that the cells which became predators were not akaryotes but microeukaryotes. Akaryotes and prokaryotes have similar bacterial organizations, and therefore similar degrees of complexity, while microeukaryotes could carry more genes, could experiment with new proteins by RNA-shuffling, and in general could express a higher degree of complexity. This was essential because complexity was the only weapon that the emerging eukaryotes could exploit against the metabolic superiority of the pro-

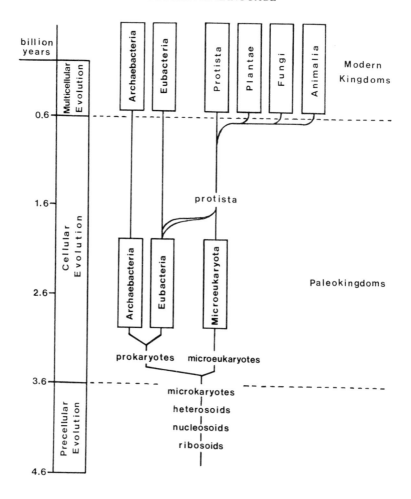

FIGURE 12:   Phylogeny of the biological kingdoms according to the Ribotype theory.

karyotes. In this respect, however, akaryotes are virtually indis-
tinguishable from prokaryotes, and I conclude therefore that
Stanier's model makes more sense in the framework of the ribotype theory than in either the prokaryotic or the akaryotic theory.

   More generally, I conclude that the ribotype theory of the
origins — which is illustrated in Figure 12 — is a valid frame-

work because it gives us a consistent set of answers to four major precambrian problems: the evolution of the ribosomes, the origin of the cell, the split of the common ancestor and the origin of the nucleus.

A NOTE ON LOCATION

A distinctive characteristic of the ribotype theory is a link between what happened before and after the origin of the cell, a *continuum* between precellular and cellular evolution. I must admit however that some developments — for example the transition from ribosoids to microkaryotes — could have taken place on another planet (Planetary and Directed Panspermia) or even in space (Cosmic Panspermia).

The ribotype theory does not require a single location and various types of intervention from space are therefore compatible with it. The atoms of our bodies, after all, were born inside giant stars, and the origin of matter belongs in any case to the Universe at large. The possibility that some developments in the history of life took place somewhere else in space would mean that the stage was wider, but not necessarily that the story was different.

CHAPTER 8

The Ribotype Theory of the Cell

COMPLEMENTARY PROBLEMS

Oparin's proposal that life originated from coacervates of proteins was first made in 1924, when proteins were thought to be responsible for all the specific characteristics of living creatures; later on, with the discovery that DNA is the molecule of heredity, a new theory proclaimed, naturally enough, that life started from naked genes. These examples show that a theory on the origins essentially reflects what biologists think about the nature of life itself, and the logic makes sense: if we understand what the cell really is, we should be able to understand, in principle, how it came into being.

The Ribotype theory of the origins has given us a new reconstruction of what happened in the beginning, and on this basis we should readdress the complementary problem of the nature of life. What is the essence of a cell? Is it really a duality of genotype and phenotype? Is the cell, in other words, a biological computer where genes and proteins are respectively its software and hardware? Is it, as some say, a throwaway survival machine built by selfish genes for their own replication?

These questions about life itself are as important as the historical problem, and in this chapter I propose to discuss the 'other' face of the Ribotype theory, the complementary side that deals with the nature of the cell and the role of its basic components.

THREE TYPES OF MOLECULES

We have seen in chapter 4 that there is a major difference between proteins and genes. In proteins, different sequences of aminoacids produce molecules which in general have different three-dimensional structures, different distributions of electric

130

charges on their surfaces, and correspondingly different biological functions. Genes on the other hand have the structure of double helices whatever the sequence of their nucleotides, have a repetitive distribution of electric charges on their external surface which makes them chemically uniform, and are suitable for only one biological function.

And the RNAs? These molecules are nucleic acids like the DNAs, and yet they behave more like proteins than genes. They are not chemically uniform and do not have a common three-dimensional structure. On the contrary, it is very easy to distinguish a transfer-RNA from a messenger-RNA, for example. As a chemical family, the ribonucleic acids have a wide variety of three-dimensional forms, different distributions of electric charges and heterogeneous functions like those that we find in transfer, messenger and ribosomal RNAs. In protein synthesis, for example, any transfer-RNA behaves as a specific catalyst, and we should not say therefore that only proteins have the ability to catalyze specific biological reactions.

Habits however die hard, and the idea that only proteins can act as biological catalysts was generally accepted until 1982, when Thomas Cech proved that RNA too behaves as a true catalyst. Some may say that Cech's result should be regarded as an exception, but this is only because the aim of his experiment was to show that RNA *alone* can act as a catalyst. In most cases RNA is combined with aminoacids and proteins, forming the compounds that I have called ribosoids, and in these systems it would be difficult to say if it is the protein or the RNA component which is responsible for catalysis. This however is a largely academic question because ribosoids behave as units and should be regarded as units.

What matters, in life, is that ribosoids are the protagonists of RNA-metabolism, that they act as true catalysts in protein syntheses, and that these syntheses are a very substantial part of all biological reactions. As a result, we can say that there are three distinct classes of specific components: genes, proteins and ribosoids.

This conclusion may seem to be a mere description of what we find in the cell, a classification based on a pragmatic inventory and not on fundamental differences, but in fact the opposite is true. There is a fundamental reason which tells us that ribosoids form a class of their own in the same sense that genes and proteins do.

We have seen that a chain of aminoacids can fold itself up in space and assume its three-dimensional form spontaneously. In order to make proteins, therefore, we only need to transfer one-dimensional information from genes to aminoacid chains, and one may think that for this process we do not need to use three-dimensional information at all. In reality, this would be true only if chains of aminoacids could be assembled in direct contact with genes, but this is physically impossible.

In order to build a protein we need molecules which are able to recognize a group of nucleotides at a certain point in space, and simultaneously to put an aminoacid in a different point where a growing chain is held in position. This requires the ability to read a linear set of instructions and to perform a three-dimensional operation at the same time, and as far as we know only ribosoids can do this.

In short, we cannot go from the one-dimensional world of the genes to the three-dimensional world of the proteins without molecules which are able to perform operations in both worlds. Notice that the crucial point, here, is not that an intermediary between genes and proteins is essential. What really matters is that this intermediary cannot consist of other genes or other proteins, because this tells us that a third class of molecules is absolutely necessary. This is why ribosoids are as fundamental as genes and proteins, and why life as we know it simply could not exist without them.

THE PERFECT AND IMPERFECT DUALITY

The idea that RNA and ribonucleoproteins are essential to protein synthesis in particular, and to life in general, was introduced into Biology by Brachet and Caspersson between the 1930s and 40s, and became universally accepted at the end of the 1950s. "It was a hard fight for me" — wrote Jean Brachet in 1981 — "to convince the biochemists that RNA is involved in protein synthesis, and this is now trivial".

Despite this recognition, however, biologists still describe the cell as a duality of genotype and phenotype, as Johannsen did in

1911. The surprising thing is that they know very well that this can only be an approximation. There is only one object in life which is a perfect duality of genotype and phenotype, and that is a virus: it carries its own genes wrapped up in a protein coat and nothing else. If the cell was a perfect duality it would be a virus-like creature, which is clearly not the case.

We can reach the same conclusion in another way. Imagine substituting the ribosomes of a cell with those from a different species: the new cell could synthesize the same proteins with the same accuracy, and its phenotype therefore would be unchanged. More in general, we could change various other ribosoids without affecting the phenotype because, by definition, the phenotype is 'the phenomenological living being', the organism as it appears to the outside world. This means that the phenotype is only a fraction of what is expressed by the genotype, that there is a hidden world in every cell that does not become visible in the phenotype. Thus biologists know very well that the duality of genotype and phenotype is not a complete picture, and yet that is how the cell is described in elementary as well as in advanced textbooks. Why? There seem to be two main reasons for this strange paradox.

On the one hand, the system which links genotype with phenotype is regarded as a mere intermediary, and has therefore a secondary importance in principle, even if in practice it is abso-lutely essential. We can call it ribotype, but a new name does not make a new theory. Unless we have genuinely new principles, and we see that the ribotype has a new role in Nature, we cannot say that we have a new theory of the cell.

The second reason is the all-powerful concept of natural selec-tion. If hereditary variations take place in the genotypes and Nature selects their effects in the phenotypes, evolution is clearly a game between genes and proteins. These are the real pro-tagonists, all the rest are made of satellites. There may be, indeed there must be, something else between them, but this does not really count. The duality is perfect and imperfect at the same time.

This gives us two main problems. Does the ribotype have a new role in Nature? And what was its real role in evolution?

THE FIRST RIBOTYPE HYPOTHESIS

One of the greatest regularities of Nature is that all eukaryotes have 80S ribosomes and all prokaryotes 70S ribosomes. Accidents of course do happen, but not of this magnitude. When they affect all creatures and divide them without exception into two neat groups we do not call them accidents any more, we call them rules. And yet, in the framework of modern biology the simple, solid, universal regularity of the ribosomes is an unexplained accident.

Let us start discussing this problem with an hypothetical question: why is there no eukaryotic cell with 70S ribosomes? As we have seen before, the molecular weight of 70S ribosomes is just over 2 million, while 80S ribosomes often exceed 4 million. By utilizing 70S ribosomes (which are equally accurate and are perfectly capable of translating eukaryotic messengers), the eukaryotes would save almost 50% of the aminoacids and nucleotides which are invested in ribonucleoproteins, often more.

This is a colossal saving of energy and material resources, and surely natural selection would have strongly favoured any mutation which reduces the metabolic burden of the cell without impairing the accuracy of translation. We also know that these mutations *must* have taken place. The mutation frequency of the ribosomal genes is so high that eukaryotes could have reduced the molecular weight of their ribosomes not just once or twice but millions of times during the history of life.

It is statistically impossible that Nature did not try the experiment, and I conclude that the experiment was indeed tried, and in all possible ways. If there is no eukaryotic cell with 70S ribosomes, my conclusion is that the experiment could not succeed: no cell could substitute 80S with 70S ribosomes and remain an eukaryote.

My interpretation consists of two assumptions:

1) There is a one-to-one correspondence between ribosomes and ribosome biogenesis. An 80S-ribogenesis can only produce 80S ribosomes, and a 70S-ribogenesis can only produce 70S ribosomes.

2) The transport of ribonucleoproteins from nucleus to cyto-

plasm is compatible only with an 80S (or higher) type of ribosome biogenesis.

These two assumptions can be put together in what I call the first ribotype hypothesis: *One Ribogenesis, One Ribotype.*

Notice that the hypothesis can be tested. By studying ribosome biogenesis in detail we should be able to explain why 70S ribosomes cannot be transported from nucleus to cytoplasm, why an 80S biogenesis never produces 70S ribosomes and why a 70S biogenesis never produces 80S ribosomes.

It should also be noticed that this is what actually happens in Nature, without exception. One cannot say therefore that the ribotype hypothesis will have an experimental basis only when the tests that I have mentioned have been performed. The experimental basis exists already, and consists of all that has been published on ribosomes and ribosome biogenesis. There is not a single fact in the literature that contradicts the hypothesis.

I assume therefore that the first ribotype hypothesis is true for two main reasons: because it is compatible with all the evidence that we have, and because it explains why all eukaryotes have 80S ribosomes and all prokaryotes 70S ribosomes. An eukaryotic cell with 70S ribosomes could not survive because it would be unable to transport ribonucleoproteins from nucleus to cytoplasm. A prokaryotic cell could produce 80S ribosomes but would be unable to compete with the most efficient prokaryotes that utilize 70S ribosomes. In this way, the first ribotype hypothesis tells us that a great regularity of life is not an accident but has a natural meaning. It makes us understand why two different types of cells have two different types of ribosomes.

THE SECOND RIBOTYPE HYPOTHESIS

Molecular phylogeny has shown that the ribosomal-RNAs appeared early in the Precambrian and have changed very little ever since. This means that some primitive microorganisms had 70S-like ribosomes while others had ribosomes of the 80S type. Let us discuss now the evolution of these cells, starting from those which had primitive 70S ribotypes.

What could they do? They could not decrease the molecular

weight of their ribosomes *and survive*, because the decrease would have made them inaccurate and they would have lost biological specificity. They could not increase substantially the ribosome molecular weight *and survive*, because such an increase would have aggravated the metabolic burden without any counter-advantage in translation, and the cells would have been eliminated by the competition.

They were condemned to keep 70S-like ribosomes, and with them a 70S type of ribosome biogenesis which is incompatible with the intracellular segregation of the genes. As a consequence, they could carry only a limited number of genes. In short, they were trapped in the prokaryotic organization. Their descendants could produce endless combinations of genotypes and pheno-types, and therefore an enormous variety of cells, but they could not change ribotype and survive. The prokaryotes started with 70S ribosomes and still have them, because it was the acquisition of a 70S ribotype that made of a cell an obligate prokaryote. In this case, the ribotype determined the cell type.

Let us discuss now the primitive microorganisms which had 80S-like ribosomes. Their potential for change was enormous, and their descendants went through spectacular transformations, giving origin to all plants, fungi and animals, and yet, like the prokaryotes, they could change virtually everything except their ribotype. If they had increased the molecular weight of their ribosomes they would have become inefficient. If they had decreased it, they would have become prokaryotes. They started with 80S-like ribosomes and still have 80S ribosomes.

We realize that the acquisition of a 70S ribotype channeled the cells into the prokaryotic line of descent, while the acquisition of an 80S ribotype created the eukaryotic line of descent. In both cases, the ribotype determined the cell type, a conclusion that represents the second hypothesis of the ribotype theory: *One Ribo-type, One Cell type.*

We see in this way a meaning in the data of molecular phylo-geny. The RNAs are the oldest molecules on earth because they are the very centre of the cell organization and came necessarily before the cell itself. Furthermore, they are the only molecules that cells are still carrying virtually unchanged within them because everything else could be changed without losing bio-logical specificity except the molecular systems that are responsible for biological specificity itself.

We have therefore a new role for the ribotype, and correspondingly a genuine new theory of the cell. A theory where the greatest discontinuity in the living world is explained at the molecular level, by reducing the dichotomy between prokaryotes and eukaryotes to the dichotomy of their ribotypes.

THE CELL AS A TRINITY

The idea that ribosomes are mere intermediaries implies that the essence of their evolution was to adapt to this role and to establish an increasingly accurate correspondence between genes and proteins. This is the concept that Woese expressed in the Progenote theory, but his biochemical evidence, on the other hand, tells us that the ribosomal RNAs changed very little during the history of life. If there is a smooth working unity between the components of the cell, and the ribosoids did not change much, clearly it was genotype and phenotype which had to adapt to the ribotype, not the other way round.

We reach the same conclusion with other ribosoids. Take ATP, for example, the small ribosoid which is universally used to exchange energy in the biological world. It would be unrealistic to say that ATP 'adapted' to processes like photosynthesis and respiration. It makes much more sense to start from ATP and say that photosynthesis and respiration evolved as ATP-based mechanisms. Another example is the genetic code, whose molecular components are all ribosoids. One could suggest that the code 'adapted' to the rest of the cell, but the idea is not very convincing; it is more likely that the rest of the cell simply adapted itself to use whatever code the ribotype made available.

In my opinion, it does not even make sense to speak of the rest of the cell as an independent entity of genotype and phenotype. 'The rest of the cell' was simply whatever managed to integrate with the primitive ribonucleoprotein systems and to form a working unity with them.

According to the ribotype theory, the ancestral ribotypes were themselves very heterogeneous systems where some ribosoids were better at carrying linear information, and others at performing three-dimensional operations in space. They had what could be called a *ribogenotype* and a *ribophenotype*, even when there was only a crude correspondence between them.

With this terminology, the origin of the cell can be summarized by saying that the genotype evolved as the extension of the ribogenotype, and the phenotype as the extension of the ribophenotype. Today, these extensions may look like the real protagonists, and the central system is declassified to the role of a mere intermediary, but this means putting first what came last and turning the logic of Nature upside down. In a way, it is like looking at the biological Universe from a Ptolemaic point of view.

We cannot get away from the fact that the ribosoids have been the essential invariants of life since the beginning. They gave to life ATP, the genetic code and protein synthesis. They created the cell in the first place and determined the cell types ever since. The cell was born as a colony of ribonucleoproteins engaged in producing other colonies of ribonucleoproteins, and still is.

THE CELL ACCORDING TO SOCIOBIOLOGY

The DNA revolution has brought us, together with genetic engineering and the model of the naked gene, a new theory on behaviour which has become known as Sociobiology (Hamilton, 1964; Wilson, 1975; Dawkins, 1976). The basic theme of Sociobiology is the genetic basis of behaviour, but here I will discuss only the sociobiological concept of the cell: the idea that cells and organisms are merely throwaway survival machines built by ruthless genes for their own replication.

There is no denying that this idea has captured the imagination of many biologists and sociologists, because it combines the classical concept of the struggle for life with the modern concept of computer technology. The idea of struggle and competition is extended from organisms and cells all the way down to the molecular world, and the DNAs are described as the molecules which emerged as victors and became rulers.

The favourite scenario for the origin of life is the theory of the naked gene, but this is not essential. The important point is that whatever started life on earth soon fell under the total control of DNA. In 'The Selfish Gene', Richard Dawkins expressed this concept with a concise statement: "Usurper or not, DNA is in

undisputed charge today". At the same time, the duality of genotype and phenotype has allowed the cell to be described as a biological computer where genotype=biological software and
phenotype=biological hardware, which means that every cell
and every organism is programmed by its own genes. Again,
DNA comes out as the ruler of the biological world.

The intuitive analogy of molecules at war, and the technological analogy of the cell as a computer, make a mixture which
appeals strongly to the intuition, and when the intuition believes
something it is usually easy to find rational arguments for it. Let
us take therefore a closer look at the two basic analogies of
Sociobiology.

In which sense can we say that DNA is a molecular usurper
which took charge of the cell? What comes to mind is an image
of DNA molecules which invaded primitive replicating systems
and enslaved them, very much like viruses which attack a cell
and force it to produce their own components. Is this analogy a
proper one? It could be if we assume that the DNA invaders and
the other molecules that were using nucleotides started competing for them and the DNAs won the contest. Personally I do
not believe in these molecular struggles, but the sociobiologists
do and should conclude therefore that DNA was the loser, not
the winner. In every cell the majority of nucleotides are devoted
to the production of ribosoids, not of genes.

In my opinion, the best parallel for the role of DNA is not the
analogy of the conqueror, but the analogy of the parasite. A
successful parasite would never drain out of its host more than a
fraction of its resources, otherwise it would risk producing their
mutual destruction. Furthermore, a successful parasite would do
something useful in return, otherwise the host might find it convenient to dispose of it.

These two characteristics fit remarkably well with the experimental pattern. The genes absorb only a fraction of the cell
nucleotides, and make themselves useful by providing stable substrates for the master copies of the biological instructions. It is
then natural to conclude that the cell adopted them and became
irreversibly DNA-dependent.

Strangely enough, the idea that some genes behave as molecular parasites was proposed by Dawkins himself (1976), and was
later developed in detail by Doolittle and Sapienza (1980) and by

Orgel and Crick (1980). The problem that these authors addressed was the paradox that large amounts of DNA in eukaryotes seem to have no phenotypic value and represent an inexplicable waste of nucleotides. They explained the paradox with the hypothesis that a colony of useful DNA is a fertile ground where a second class of parasitic DNA can spread and replicate. No other explanation for its existence is required other than its natural tendency to parasitize other genes.

This second class of DNA was referred to as *selfish DNA*, and Orgel and Crick labeled the selfish genes as "the ultimate parasites". The solution is simple, elegant and logical, but why go only half way? The selfish genes were the parasites of useful genes as these had been the parasites of previous RNA-genes. In this way we have one mechanism for the origin of all genes, which means that all DNAs have the same status as molecular parasites.

Let us now return to the statement that "DNA is in undisputed charge today". If this simply meant that DNA is the carrier of genetic information, all biologists would accept it, but the sociobiologists have much more than that in mind. Their meaning is that the cell economy is entirely enslaved to the reproduction of DNA, and every process in life is subordinated to that primary goal. This is the meaning of the definition of every organism as "a throwaway survival machine" built and used by the genes as a disposable container, and it is this second interpretation that I find illogical.

If DNA is a molecular parasite, we cannot say that it is in charge of the cell any more than a parasite is in charge of its host. Nor can we say that DNA determined the origin and the evolution of life any more than we can say that the evolution of a parasite preceeded and determined the evolution of its host.

The Ribotype theory puts the ribotype metabolism at the center of the cell economy in the same sense that we attribute to the metabolism of a host a conceptual and an evolutionary priority over the metabolism of its parasite, without underestimating that the original parasite has become an essential part of the system.

I would like to emphasize that the Ribotype theory gives a secondary role to DNA only during precellular evolution. From the end of that period, the theory describes the cell as a trinity of genotype, ribotype and phenotype, where every member is conceptually as important as the other two otherwise we would have no trinity at all. In this sense, the importance of DNA is undiminished. With the Ribotype theory, in short, all that we lose is the mythology of DNA, not the perception of its real role in Nature.

COMPUTERS, VILLAGES AND CITIES

The classic duality of genotype and phenotype, as we have seen, has been used to describe the cell as a biological computer and to conclude that every living creature is programmed by its own genes. If we want to use the analogy, however, we should use it properly and take into account that in addition to software and hardware there is an essential third party in the computer world: the human beings who built the hardware and wrote the software.

For practical purposes we can of course describe the computers as dualistic objects, but we cannot understand their origin and evolution with two categories only. In the same way, we cannot understand the origin and the evolution of the cell if we forget that the ribotype was the all-important third party which built genotype and phenotype. We would have a chicken-and-the-egg paradox which is as absurd as the idea that computers evolved either from a primitive software or from a primitive hardware.

Some readers may not like analogies but, rightly or wrongly, these are widely used to illustrate scientific concepts, and I would like therefore to add another one to the list. It is a story about villages and cities with which I concluded my first paper on the Ribotype theory in 1981, and it goes like this.

A small cell contains 5 or 10 thousand ribosomes, the population of a village, while big cells have millions of ribosomes, like the inhabitants of large cities. Imagine now asking yourself about the origin of these communities. The first step is fairly obvious: the big cities were once small villages that just grew bigger. The equivalent is the theory that eukaryotes derived from prokaryotes, a most sensible proposition.

But what about the first villages? Here the situation becomes more complicated because we have two distinct possibilities. One view is that all the different building blocks which make up the objects of the village had the ability to aggregate and produce rudimentary houses, rudimentary machines and rudimentary inhabitants.

The natural evolution of houses, beds, chairs, typewriters, TV-sets and bicycles eventually produced inhabitants that were more and more able to use them. They developed eyes to watch TV, fingers to type, legs to ride bicycles and so on. The inhabitants adapted to the evolving objects in the village and eventually achieved a perfect integration with them.

The second view states that all this is nonsense. The real things are the books. It is they which contain the instructions for making all the other objects, and these objects, including of course the inhabitants, were built with the specific purpose of producing more books. The buildings were made from drawings of the buildings to produce in them other drawings, the telephones in order to produce telephone directories, the inhabitants in order to produce books of anatomy, and so on.

In a fit of madness, somebody comes out with a third idea, and says that it was the inhabitants who built villages and cities, but this is promptly rejected. Everybody can see that in a city the inhabitants are going around busily making objects, and are obviously the instruments with which the city keeps itself going. On the other hand, how could the inhabitants produce buildings without drawings, streets without maps, how could they do anything without textbooks, dictionaries, memos and guidelines? The suggestion is patently absurd. Its main argument is that only the inhabitants can read the instructions in the books and build objects from them. But who is going to believe it?

DIFFERENTIATION AND PRERIBOSOMES

Any theory on the nature of the cell has implications for our approach to what is probably the greatest mystery of multicellular life. This is differentiation, the process by which the cells of a developing embryo give origin, with billions of operations finely coordinated in space and time, to the various tissues and organs of the body.

The most popular view of differentiation, today, is the theory of differential gene expression, which reduces the process to the switching-on of certain genes and the switching-off of others. Many biologists, however, have pointed out that this theory tells us how differentiated cells use the genetic information, not how they became differentiated in the first place. Differential gene expression may well be only the consequence of differentiation, not the cause of it. After all, in eukaryotes (the only organisms which are capable of true differentiation) transcription is not followed immediately by translation, and in some cases genes can be transcribed and their messengers never translated.

Whatever is the case, biologists have found it necessary to divide the control mechanisms of protein synthesis into two major classes: mechanisms that switch genes on and off (genetic or transcriptional control), and mechanisms that operate at later stages (post-transcriptional control).

Since the post-transcription mechanisms affect the ribotype, I prefer to call them 'ribotype control mechanisms', and this is not just a question of terminology because, in principle, a ribotype control can operate both before and after transcription. The ribotype control mechanisms form therefore a potentially wider class for which we have at least the evidence of the post-transcription mechanisms, whose existence is very well documented and whose importance is widely recognized.

Let us concentrate now on the particular subclass formed by the ribotype control mechanisms that regulate the biogenesis of ribosome precursors and their transport from nucleus to cytoplasm. Let us call them 'preribosome control mechanisms', or mechanisms that regulate the activity of the 'preribosome system', where 'preribosomes' means ribosome precursors (from the beginning of biogenesis to complete maturation), and the preribosome system is the subcellular apparatus formed by all the preribosomes of an eukaryotic cell. Using this definition, the preribosome system begins with the nucleolus, continues in the nucleus with a network of ribonucleoproteins, and ends in the cytoplasm wherever newly formed mature ribosomes are engaged for the first time in protein synthesis.

While the genetic system is the source of messages, the preribosome system sends to the cytoplasm the machines that translate these messages: we can look at the cell as a factory which produces not only certain goods, but also the machines that make these goods. Since the machines deteriorate after a natural working cycle, we expect that the production of the machines is a function of their working lifetime and of the overall working load of the factory, i.e. the greater the cell activity in protein synthesis, the higher should be the production of preribosomes.

The cell, however, does not seem to follow this sound industrial logic. Embryonic cells have a very high turnover of preribosomes in respect to their protein syntheses, far greater than anything which normally happens in adult life. Well-differentiated cells, on the other hand, can be very active in protein synthesis with only a modest stream of precursors to replenish their cytoplasmic stock of ribosomes (red blood cells can even produce hemoglobin with their reservoir of ribosome precursors completely cut off).

There is, in other words, a discrepancy between preribosome-metabolism and ribosome-metabolism, but let us take a closer look: the discrepancy exists only if we assume that the *exclusive* role of the preribosome system is to produce ribosomes. If Nature had another role for it and the two roles followed different patterns, we would no longer expect to observe one pattern, but a combination of the two. There is therefore a very natural meaning in the observed experimental pattern: the cell must use the preribosome system for producing ribosomes and for *something else*.

I have been trying to find this something else with a variety of experiments with chick embryos, and my preliminary conclusion (published in 1974 and 1983) is that the preribosome system is a sort of 'pacemaker' for the cell. It seems to have a natural cycle which ends by triggering mitosis and which is gradually prolonged during differentiation by the imposition of constraints (the preribosome theory).

A similar conclusion was reached by Robert Drummond in 1980 (for a short review see the paper by David Horrobin in 1981) with totally different methods, and in my opinion this convergence is significant. We are perhaps beginning to see a meaning in the fact (discovered by Jean Brachet in the 1940s) that the morphogenetic gradients of a developing embryo are gradients of RNAs and ribonucleoproteins. Another intriguing correlation, suggested by the students of Virchow, is that "the only constant feature of neoplasia was nucleolar aberration in morphology and in nucleolar function" (quoted from Bush and Smetana, 1970).

For the time being, at any rate, what matters most in this field is that a very large body of experimental facts accumulated by many scientists, has already proved the existence and the importance of various ribotype control mechanisms in differentiation.

Opinions still differ, and at present we do not have a theory of differentiation which accounts for all known facts. My preribosome theory, for example, is not only questionable but necessarily incomplete, because it concentrates on ribosome pre-cursors and leaves out important ribotype control mechanisms, like those which regulate the metabolism of transfer-RNAs, messenger-RNAs and mature ribosomes. Differentiation, in short, is still a mystery, but at least one message has emerged: future generations will have to determine not 'if' but 'how much' the ribotype is important in it.

THE CELL AND NATURAL SELECTION

The fact that differentiation is still a mystery means that the last of the great precambrian events — the origin of multicellularity — is a problem for which we have no solution. This is regret-table, but on the other hand we must not confuse the historical reconstruction of what happened in the past with the study of the mechanism of evolution which operated in all periods.

We all believe (I presume) that there are invariant laws of Nature, and we are also prepared to accept that there might have been an invariant mechanism in evolution. But is natural selec-tion such a mechanism? This is a problem that we cannot dis-sociate from the complementary problem of the nature of the cell.

We have seen that natural selection is one of the major reasons for which biologists keep describing the cell as a duality of geno-type and phenotype. Alternatively, we can start from the classic picture of the cell and conclude that the mechanism of evolution is a duality of variation in the genotypes and selection in the phenotypes (the combination of the two processes is natural selection, which acts on the genotypes through the phenotypes). The nature of the cell and the mechanism of evolution are clearly two faces of one problem.

This means that if the cell is a trinity of genotype, ribotype and phenotype, there should be a mechanism of evolution which is more general than natural selection as the trinity is more general than the duality. A few years ago, I was convinced that we would have to understand differentiation before we could start looking for a more general mechanism of evolution, but later I changed my mind. If Darwin discovered natural selection without know-ing anything about Mendelian genetics and molecular biology, perhaps we should be able to generalize natural selection even if we do not know how differentiation works. It should not be a question of collecting more facts, but of looking at the facts that we already have from a different angle.

From an historical point of view, the Ribotype theory of the Origins does not go beyond the Precambrian because differentiation is at present an unsurmounted obstacle. From a biological point of view, however, the Ribotype theory of the cell tells us that we should start looking now for a more general mechanism of evolution. This is the problem to which I will dedicate the last chapter of the book.

CHAPTER 9

The Semantic Theory of Evolution

THE CONCEPT OF ADAPTATION

Adaptation is one of the oldest concepts in Biology and a very real phenomenon. Fishes are clearly adapted to swimming, birds to flying and worms to burrowing. Each creature seems to have been designed for its life style, and a design suggests a Designer: biological adaptation was probably as important for Religion as it was later for the natural sciences.

Darwin identified adaptation with evolution itself. In fact, he did not much like the word 'evolution' because it implies a trend towards progress which is not universal: some creatures remained virtually unchanged for eons, while others actually regressed. According to Darwin, we should speak not of 'evolution' but of 'adaptation by natural selection', because that much better describes what happens in the history of life.

The environment's resources are limited, and every population has a reproductive potential which is far greater than its actual number of descendants; as a result, the individuals which are better adapted to the environment (the fittest) have more chance to benefit from its resources, and are likely to leave more descendants. A change in the environment, on the other hand, does not affect all creatures in the same way; it can be good for some and bad for others. As a consequence, the successful organisms of any historical period are not necessarily better than their predecessors, but they are always better than their *coexisting* competitors. This is why the history of life, according to Darwin, is a history of adaptations.

If we take into account that the environment changed locally and globally with time, we understand why some creatures improved, others deteriorated, and some did not change at all. We also understand, in Darwin's view, why all present-day creatures are to a greater or lesser extent adapted to the environment. But do we?

Thomas Hunt Morgan, at the beginning of our century, was probably the first person to suggest that Darwin's argument explains nothing because it is a circularity. Which creatures survive? Those that are better adapted to the environment. And which are better adapted? Those that survive. The 'survival of the fittest' seems to be nothing more than the 'survival of the survivors'. Let us discuss this point with an example.

A team of scientists brought two types of flies, one with long wings and the other with short ones, to an island that was constantly swept by strong winds, and observed what happened. There were two possibilities. If the long-winged flies survived, they would be the fittest, presumably because long wings give them a more powerful defence against the wind. If the short-winged flies survived, on the other hand, they would be the fittest because they offer a smaller surface to the wind and are less likely to be carried away.

As it happened, the short-winged flies turned out to be the survivors, but for our purpose the outcome is irrelevant. The important point is that whichever type was going to survive would have been the fittest anyway, which shows that fitness and survival are merely two different words for the same thing.

One could argue however that the experiment was not done properly. Those scientists should have studied the physiology of flight beforehand and determined in advance which type is objectively the fittest to cope with strong winds. Then they should have performed the experiment on the island to test if the fittest was actually the type that survived.

This brings us to the core of the problem: if 'fitness' and 'survival' can be defined independently, the 'survival of the fittest' is a true scientific concept, and can be tested both in the laboratory and in the wild. Otherwise it is a circularity and explains nothing. (A similar situation exists in Physics. Newton's second law, $f=ma$, is a circularity if force is defined in terms of mass and acceleration only. To avoid this, physicists must find an independent definition and say, for example, that force is the derivative of the potential). Clearly, we have to examine the concept of adaptation a little more closely.

THE RED QUEEN

In 'The Development of Darwin's Theory' (1981) Dov Ospovat has shown that there were two main phases in Darwin's approach to adaptation. The first lasted from 1837 to the essay of 1844; the second from 1854 onwards.

According to Ospovat, in 1844

Darwin believed that species were perfectly adapted, that adaptation was mainly to the physical environment, that perfectly adapted species did not vary, and that species varied only when the environment changed. Evolution occurred when species became imperfectly adapted because of some environmental change. The species would then start to vary, and natural selection would favour the best adapted variants until some new state of perfect adaptation was reached.

From 1854 onwards, this picture is dramatically changed.

Species are now continually struggling and competing against each other, continually varying and continually evolving. Adaptation, now conceived as mainly to the biotic environment, is no longer perfect. Species were only as relatively good as their competitors.

Today, this second Darwinian concept of adaptation, "imperfect and dynamic", is accepted by most biologists. Let us see why.

In 1913, L.J. Henderson published 'The Fitness of the Environment', in which he pointed out that the environment too has to be fit for life, and not just life for the environment.

Darwinian fitness is a mutual relationship between the organism and the environment. Of this, fitness of the environment is quite as essential a component as the fitness which arises in the process of organic evolution.

In addition to this, biologists have become increasingly aware that organisms not only adapt to their local environment, or 'ecological niche', but often modify it, create new niches and have a direct role in the environmental changes that take place around them. There are organisms that live inside other organisms, for example, and the transformation of the whole earth into an oxygenated planet was brought about by microorganisms. Trees do not just live in the soil, but remake it. Termites, ants and bees build nests and hives which are the very centre of their ecological niche.

These examples and a multitude of similar cases have proved beyond doubt that the interdependence of organisms and niches is an experimental reality. And there is also no doubt that both changed in the past. The history of life therefore was written by the evolution of organisms as much as by the evolution of niches, but there is a substantial difference between the two processes. We can say that organisms adapted to niches, but much less that niches adapted to organisms. The production of oxygen, for example, was not a phenomenon by which the atmosphere 'adapted' to life. Oxygen was a poison at the beginning, and if the microorganisms had not changed drastically to cope with it, it would have remained a poison.

The history of niches, although modified by organisms, was essentially a history of runaway processes and random changes. We can see that the organisms had a goal and a direction: they had to adapt to the environment and keep following it wherever it was going. But where was the environment going, in the Darwinian framework? Apparently nowhere. The environment as a whole just keeps changing, while the earth goes in circles around the sun.

This is the Darwinian concept and this, according to most biologists, is the modern idea of adaptation and evolution. "The modern view of adaptation" — wrote Richard Lewontin in 1978 — "is that the external world sets certain 'problems' that organisms need to 'solve', and that evolution by means of natural selection is the mechanism for creating these solutions".

Van Valen has recently reproposed the second Darwinian concept of adaptation and has named it 'The Red Queen hypothesis', after the character created by Lewis Carroll. In 'Through the Looking-Glass', the Red Queen says to Alice: "Here, you see, it takes all the running you can do, to keep in the same place".

This was true for Darwin and for many biologists it remains true today, which explains why Darwinism is so strong. Adaptation is a paramount reality of life, and as long as we accept the Darwinian concept of adaptation Biology is bound to have a Darwinian heart. But how good is the Red Queen hypothesis?

ORGANISMS AND NICHES

"To a jawless fish in the Ordovician" — wrote George Gaylord Simpson — "the idea of land life would have been ridiculous: there was no food on land". Why should one of its amphibian cousins have developed a complicated respiratory system when it could have stayed in the sea? If there had been "intermediate niches" between the water and the land, we could imagine some creatures venturing from one niche to the next, and in time coming out on dry land altogether, but that was not the case. The seashore drew the line between two drastically different worlds. And what made a reptile alter the bones, the muscles and the skin of its forelimbs to produce wings and take to the air? Again, there is no intermediate niche between the land and the atmosphere.

The Darwinian idea that organisms keep tracking the environment in a never-ending chase accounts for gradual transformations, but not for dramatic events like the occupation of the land from the water and of the air from the land. For these transitions, Darwin's theory requires not only intermediate varieties that *perhaps* did not exist, but also intermediate niches that *certainly* did not exist.

But adaptation is a reality. Amphibians did invade the land, and reptiles did develop wings which took them into the air. It is not adaptation which is at fault, it is the Darwinian theory that is behind it. In addition to the paleontological argument there is also a theoretical objection to the Darwinian concept of adaptation.

Organisms and niches are so interdependent that we cannot define a realistic ecological niche without the organisms that occupy it. When we try, we often end up with niches that should be occupied but that are not. Snakes live in the grass, for example, and birds make nests in trees, and yet no snake eats grass and no bird eats leaves, despite the fact that if they had adapted to these diets they would have had an endless supply of food. Furthermore, if we separate organisms and niches we could not account for a formidable number of niches that exist only because organisms made them.

This creates quite a problem. If niches can only be defined by the organisms, we start inevitably with a perfect fit between them, but in this way there is no space left for adaptation. We have to assume therefore that adaptation is perfect and imperfect at the same time: perfect in our definitions and imperfect in Nature. It is the only way of explaining why we have both the illusion that organisms and niches form a perfect fit, and the impression that they never do.

The strict interdependence of organisms and niches means also that the 'survival of the fittest' is a tautological concept. We have seen that the circularity can be avoided only if fitness and survival are defined separately, but this separation is impossible if niches can only be defined by the organisms that occupy them. This is why Ronald Fisher concluded, in 'The Genetical Theory of Natural Selection' (1930), that 'the sole criterion for fitness is the number of offspring''. The Darwinian concept of adaptation, in short, does not account for the great novelties in the history of life and amounts to saying that 'survivors survive'. Clearly, we need something better.

ENVIRONMENT AND NATURAL CYCLES

Let us go back to the Precambrian and re-examine some of its events from the point of view of adaptation. We have seen in chapter 2 that there are three types of creatures in Nature (producers, reducers and consumers), and that the first cells were definitely anaerobes, probably heterotrophs and possibly fermenters. Let us start from this scenario and see what could have happened.

On the one hand, the first cells were likely to reproduce with a great number of mistakes, which means that they expanded slowly (most errors are lethal) and that their descendants became increasingly different. On the other hand, we have seen that precellular evolution produced a wide range of organic molecules, which means that different cells could survive on different diets. Biological diversity, in other words, could be created by hereditary variations and, once in existence, could be maintained by the diversity of the food supply. Let us follow the further evolution of this diversity.

The organic waste of the primitive fermenters was constantly increasing the reservoir of small organic molecules and, at the same time, the dead cells of past generations were also becoming an increasing reservoir of organic matter. The further diversification of the ancestral consumers could be channeled either in the direction of dismantling dead bodies (like fungi do), or in the direction of utilizing smaller and smaller organic molecules. Both developments required an extension of the metabolic machinery, and we have seen that an extended fermentation provides steps that lead either towards increased catalysis and pre-respiration or, in reverse, towards photosynthesis. This gave some ancestral consumers the means of following two diverging lines of development: one towards the first reducers and the other towards the first photosynthesizers.

There is evidence from the fossil record that photosynthesis appeared about 3 billion years ago, in the first half of the Archeozoic, long before the beginning of the oxygen revolution. This means that a complete system of producers, consumers and reducers came into existence when the environment was still anaerobic. Notice that there are various bacteria which phosynthesize without releasing free oxygen, and the oxygen revolution therefore was neither a necessary nor an inevitable consequence of the invention of photosynthesis. Life could well have gone on indefinitely at the anaerobic level, since the chain of anaerobic producers, consumers and reducers was self-supporting and therefore potentially immortal.

But the blue bacteria did appear, oxygen began accumulating, and an enormously slow oxygenation of the planet took place. Most microorganisms adapted to the change, and approximately a thousand million years later, a new self-supporting and potentially immortal chain of aerobic producers, consumers and reducers had replaced the old anaerobic one in almost every corner of the planet.

At the same time, another type of cellular engineering was taking place, and at the end of the oxygen revolution the first truly eukaryotic cells — the protista — appeared. Their predecessors had managed to survive by becoming predators, and as predators they expanded everywhere, but then something else happened. Instead of remaining predators, the protista differentiated into eukaryotic producers, consumers and reducers, i.e. into all the cell types that make up a self-perpetuating metabolic chain. And when multicellular organisms appeared — 6 or 7 hundred million years later — the same process took place all over again: they diversified in multicellular producers (plants), multicellular consumers (animals) and multicellular reducers (fungi).

Notice that this pattern is not an *a priori* necessity for survival. The monocellular eukaryotes could have easily survived by remaining predators, because the prokaryotes provided a stable supply of food. And the same is true for the multicellular organisms. Their ancestors fed on single cells, and their descendants could have survived on that diet indefinitely, as some of them still do. And yet they started from scratch all over again and built a totally new self-perpetuating system.

One can feel that we are in the presence of a pattern, here, of a great pattern of Nature. As soon as a new type of cell organization is developed, the same phenomenon occurs: the organisms differentiate into all the various forms that create a self-perpetuating cycle. As if, in the long run, the only creatures that make sense are those that become members of natural cycles. Shouldn't we say that organisms adapt not to the environment, but to the natural cycles?

Consider now the environments, or the various ecological niches where organisms live. When we draw a diagram of the cycle of oxygen, for example, we treat the atmosphere as a member of the cycle, exactly as we do with the organisms. Being a member of a cycle simply means receiving something at one end and delivering something else at the other end, and in this respect organisms and niches behave equally. Instead of saying that organisms adapt to niches, therefore, we could say that organisms and niches adapt to the natural cycles, and in this way give origin to self-perpetuating chains. This is the alternative to the Darwinian concept of adaptation that is taking shape: instead of 'Adaptation to the Environment' we are beginning to speak of 'Adaptation to Natural Cycles'.

THREE OBJECTIONS

The concept of natural cycles is very old, and so is the concept of adaptation, and yet the two have not been put together. Biologists have repeatedly spoken of adaptation to the environment, but not of adaptation to natural cycles. Given the rise of ecology in our century, this is surprising, but perhaps three objections played a part.

One argument is that natural cycles are generally associated with absolute necessities. It is obvious that life needed them, but a necessity is not something that we can speak about in terms of adaptation: it would be like saying that life adapted to the laws of physics and chemistry. This deterministic view however does not stand up to a close scrutiny. The oxygen cycle, for example, was not a necessity since life could well have gone on indefinitely at the anaerobic level. Many cycles disappeared when some species became extinct, and yet life went on. The cycles of life were created, changed and some died out: they were historical developments, inventions of Nature, and not timeless rules like the laws of physics.

A second argument is that organisms adapt to the environment, but the environment does not adapt to anything because it is shaped by uncontrollable forces. Organisms adapt because they are centres of biological order, while the environment is the domain of physical disorder. Earthquakes, volcanic eruptions, storms and lightning strike at random, and invariably lead to chaos and destruction.

This objection does not take into account that genetic mutations also strike at random, and in most cases are as lethal as physical calamities. As for the environment, one should remember that there are also great regularities in it, like the cycle of day and night and the constant rhythm of the seasons. There is a considerable amount of disorder in organisms and of order in their ecological niches, and we cannot say therefore that only organisms could adapt. All we can say is that in general niches did not adapt to organisms, and we should conclude therefore that both organisms and niches adapted to 'something else'.

A third objection is that a natural cycle is made of organisms and ecological niches, and the idea of adaptation to a cycle therefore can be reduced to individual adaptations of organisms to niches. This argument forgets that biological diversity is created at random, and it is statistically impossible that all organisms which appeared on earth actually managed to form natural cycles. Only a few did, but in the long run only these few survived, and we are left therefore with the impression that adapting to the environment also means adapting to a natural cycle, while in fact the two processes are largely independent.

BEYOND THE PARADOXES

We have seen that the classical concept of adaptation to the environment introduced into Biology a circularity and a paradox: the circularity that survivors survive, and the paradox that adaptation is perfect and imperfect at the same time.

The circularity arises because organisms and niches are so interdependent that we cannot distinguish the creatures that survive from those that are best adapted to their niches. We have seen however that being adapted to a niche does not also mean being adapted to a natural cycle, and in the long run only the self-perpetuating chains of the natural cycles ensure survival. With the new concept of adaptation, therefore, the circularity disappears because the survivors are not always and not necessarily the organisms that are best adapted to their niches.

Let us now turn to the paradox. Again, the origin of the trouble is that we can only define organisms and niches together, as if they were perfectly adapted, and yet they cannot be perfectly adapted because in this case there would be no evolution. Richard Lewontin put it in this way:

If ecological niches can be specified only by the organisms that occupy them, evolution cannot be described as a process of adaptation because all organisms are already adapted.

But let us look at the new concept. Imagine an emerging natural cycle in which only one position is not occupied. An organism which can fill that position would complete the cycle, and it would not matter if the creature is not well adapted to its niche. The advantage of being part of a cycle would benefit that organism far more than being adapted to a niche. In the new framework, in other words, the idea of perfect adaptation to a niche does not appear at all, even in principle, and the paradox simply disappears. We can say therefore that evolution is indeed "a process of adaptation", but primarily adaptation to natural cycles, not to the environment.

In Paleontology, the Darwinian concept implied the existence of intermediate niches for the great transitions of life from water to land and from land to air, while the new concept does not. The oxygen released by microorganisms in the sea can be used by creatures on the land which have never been anywhere near the sea. The natural cycles are networks which effectively link organisms and niches beyond the physical barriers of space and time.

Perhaps they are not as easy to visualize as real organisms and real niches are, but the history of science has always been a struggle to go beyond appearances and reach for the reality of the invisible. The concept of adaptation to natural cycles may look like an abstraction, but it also looks like the essence of evolution. Before we can say that this is really the case, however, we first have to discuss natural selection. This is the all-important mechanism of the Classical theory, and the major stumbling block for any new theory of evolution.

THE EVOLUTION OF LANGUAGES

The suggestion that man descended from ape-like ancestors caused a scandal, but nobody seems to have objected to the idea that languages like French, Italian and Spanish 'evolved' from Latin. Perhaps it was taken for granted that linguistic evolution has nothing in common with biological evolution, but such an assumption was probably premature. There are certain parallels between the two processes that deserve to be examined with great care.

The modern theory of information was developed from Thermodynamics and Electromagnetism by generalizing some of their concepts and expanding their range of application; the formulae that describe how many messages can be sent through an electric cable, for example, have been used for a variety of other signals. A computer scientist speaking about the information of a message, nowadays, does not have any particular physical carrier in mind. It could as well be the information of molecules and microchips as of books and conversations. There are formal properties in the exchange of messages that just happen to be independent of the systems that carry them. It is not absurd therefore to think that there may be a useful parallel between the evolution of languages and that of organisms.

There is also another point in favour of the analogy. In three university courses held in Geneva between 1908 and 1911, Ferdinand de Saussure made the same generalization about linguistics that Johannsen made about biology in those very same years. He proposed a fundamental distinction between *langue* and *parole* which is equivalent to the distinction between genotype and phenotype. 'Langue' is the linguistic genotype, the pool of signs that are available to all users of a language, the equivalent of the pool of genes in a population. 'Parole' is the linguistic phenotype, the actual use of words and signs that individuals make when they communicate.

For a mathematician, the fact that we can speak of genotype and phenotype in both organisms and languages means that the parallel between their evolutions is perfectly valid, because he would concentrate on the characteristics of the processes that do not depend upon their physical carriers. A biologist however would be more cautious, and would first examine a variety of objections.

One is that languages were developed by creatures who can anticipate the future and plan ahead. The parallel would seem therefore to imply that behind biological evolution there was an Intelligence, as intelligent beings were behind the evolution of languages. But let us examine this argument a little more closely. The fact that languages were developed by human beings does not mean that they were the result of a *conscious* activity. For one thing, there are various forms of communication in animals which show that the ability to exchange messages does not depend upon the ability to anticipate the future. Secondly, all forms of conscious manipulation of words and signs almost certainly came *after* the languages, and not before.

The role of consciousness in the evolution of languages was rather like the role of artificial selection in the animals that we see around us. If the intelligence that was present in the breeding experiments does not prevent us from talking about natural selection, we should not be deceived by the fact that there was also some conscious activity in the history of languages. The important point is that this was secondary and came too late anyway.

We conclude therefore that the argument from consciousness

is not valid. The evolution of languages was a natural process, not the result of a conscious activity, and the parallel with the natural evolution of organisms cannot be ruled out on this ground. There are, however, other objections to take into account.

INFORMATION AND MEANING

Let us turn our attention to the mechanisms by which languages evolved. Random mistakes (like copying errors) were certainly made and were followed by a selection which perpetuated some and discarded others. This is a two-step process that we can rightly call *linguistic selection*. There was therefore in the evolution of languages a mechanism that is perfectly equivalent to natural selection, but it is easy to realize that such a mechanism had only a secondary role.

We cannot say that letters and words were created at random and then people selected the combinations that made sense. A word can have any meaning you like. The word 'mare', for example, means 'horse' in English and 'sea' in Italian. The information content of the word 'mare' is exactly the same because to create it we have to choose the same letters and arrange them in the same order in both languages. The meaning of that word however is profoundly different because it is created by a *linguistic convention* which has nothing to do with the instructions necessary to create the word itself.

Information can exist without meaning (with the appropriate instructions one can write any word without knowing what it means), but meaning cannot exist without information. A language, on the other hand, was not developed by people sitting around a table and deciding what meaning to give to each word. They could not have done so without a language anyway. We cannot say that meanings were created by a conscious activity, at least at the beginning, but we also cannot say that a language evolved without meanings. We conclude therefore that there must have been a natural mechanism of linguistic conventions that somehow *created* meanings and made languages evolve. In 1972, in fact, Noam Chomsky presented evidence for an instinctive or unlearned capacity for generating meaningful sentences.

We have in this way two distinct mechanisms: the evolution of languages took place by linguistic selection and by linguistic conventions, and this last mechanism was by far the most important. We have already seen that natural selection is the equivalent of linguistic selection, but do we also have in Biology something equivalent to the linguistic conventions? Do we have biological processes which add natural meaning to information? We do, and since they are perfectly equivalent to the linguistic conventions I will call them *natural conventions*.

The conventions that established the genetic code, and those that led to the choice of right-handed sugars and left-handed aminoacids are the best known examples, and show that natural conventions not only existed but had a very important role. We can say therefore that there were two distinct mechanisms also in the history of life: biological evolution took place not only by natural selection but also by natural conventions.

At this point however some may disagree. The origin of the genetic code was a convention in the sense that Nature could have chosen other codes, but it was also a "frozen accident". The choice of right-handed sugars was also arbitrary, but again it was an accident, as was the choice of left-handed aminoacids. Can we really say that a few isolated accidents represent a distinct mechanism of evolution? Natural selection is perfectly compatible with random accidents of this kind, and there is no need to invent a new mechanism for them.

So it would seem, but we should not jump to conclusions. Nobody has actually shown that the genetic code was an isolated accident, and we should not take this for granted. As for the existence of a qualitative distinction between conventions and selection, we know that in Linguistics this distinction is real and that it is also very real in Information theory and Computer science. Before saying that there was no such distinction in biological evolution, we should at least take a closer look at it.

A NEW LAW OF THERMODYNAMICS

The two laws of classic Thermodynamics, the laws of energy and entropy, have given us a formula that describes the information content of a natural system. Computer science and Information theory have used this formula extensively, but have concentrated on information only and have systematically avoided the concept of meaning. Recently however things have begun to change. Lila Gatlin, for example, has shown that a technicality called redundancy can be described as a n-dimensional vector which may be used as a measure of *thermodynamic meaning*, or *meaningful information*, or complexity in the Von Neumann sense:

quantitative measures for the characterization of the meaning of a message are now emerging (Gatlin 1977).

There is still a long way to go in this direction, but we can already see the overall implication of these developments. If processes which add thermodynamic meaning to thermodynamic information exist, then it is obvious that something is missing in our description of Nature. In addition to the laws of energy and entropy, we need a third law that accounts for the existence of meaningful information.

We have not yet found a general mathematical expression for it, but the essence of the third law can be described, like the other two, in simple words. The first law states that: "energy can neither be created nor destroyed". The second law that: "global entropy can be created but not destroyed". The third law that: 'meaningful information can both be created and destroyed". Lila Gatlin has represented them with the scheme of Figure 13, which illustrates that energy, entropy and meaning are thermodynamically autonomous and logically complementary to one another.

In this way we can divide all physical processes into two great classes: one for the reactions which are completely described in terms of energy and entropy only, and the other for those where thermodynamic meaning is created or destroyed. In this second class, we can regard as a natural convention whatever process is responsible for the exchange of meaningful information, and we obtain in this way a qualitative but rigorous definition: "a natural convention is a process where thermodynamic meaning is either created or destroyed".

If we had a mathematical formula for the third law, the natural conventions — the biological equivalent of the linguistic conventions — would be described by equations as ordinary

LOGICAL PATTERN OF BASIC LAWS

	Created	Destroyed
Energy	–	–
Global Entropy	+	–
Meaningful Information	+	+

FIGURE 13:   Schematic illustration of the laws of thermodynamics (from Lila Gatlin, 1977).

chemical reactions are, and there would be no doubts about their objective reality. But let us ask ourselves this question: do we really need a mathematical formula to be convinced that natural conventions are real thermodynamic processes?

Look at the sugars, for example. Inorganic reactions produce right-handed and left-handed sugars in equal amounts, and there is nothing which makes one type better than the other. And yet 'something' happened in the history of life, and after that all biochemical reactions produced and used only right-handed sugars. That 'something' was very real, and is a perfect example of a natural convention, of a process that 'makes a choice' and channels all reactions in one direction only.

The important point is that that choice was not the result of selection. We cannot say that some creatures appeared with right-handed sugars and others with left-handed sugars, that they competed and that eventually the right-handed won the contest, because for all we know right- and left-handed sugars perform equally well. Natural conventions, in other words, are qualitatively different from natural selection. We conclude that the distinction between them is justified from a thermodynamic point of view, and is documented by historical examples. This means that biological evolution, like the evolution of languages, could and did exploit two distinct mechanisms.

NON-DARWINIAN PROCESSES

Let us go back to the origin of life and look at the ribotype reconstruction from the point of view of natural selection. The reader will have noticed that in chapter 7 I have not described the evolution from ribosoids to cells in Darwinian terms. I have not said, for example, that nucleosoids competed among themselves, that the fittest were favoured by natural selection, and that there was a struggle for life which led to a gradual increase in complexity all the way up to the first cells.

I have simply said that certain objects appeared because their natural properties allowed them to do so, in the same way that from hydrogen and oxygen we see that in certain conditions water appears. Ribosoids, nucleosoids and heterosoids were enormously heterogeneous, and I have repeatedly pointed out that they appeared with percentages and frequencies that were determined by statistical factors only.

Do we need to assume that these percentages and frequencies were also influenced by processes like competition and natural selection? We could, but we do not have to. Let us assume, for example, that nucleosoids replaced ribosoids, that heterosoids wiped out the lower nucleosoids, and that heterosoids with membranes and DNA led all the others to extinction and became the first cells. With this scheme we end up with cells in an environment which is very poor in organic molecules, and it is difficult to justify their survival, particularly because the first cells were probably heterotrophic creatures that needed a wide variety of nutrients.

If we assume instead that there was no competition, we have a scheme where many different systems could go on undisturbed producing descendants by quasi-replication, and the simplest ribosoids worked side by side with the most complex heterosoid, like bacteria today are undisturbed by large mammals. In this framework the first cells would appear in solutions that are rich in all sorts of nutrients and this is (a posteriori) a very desirable characteristic.

I am not saying that what looks good was also necessary. I am simply very reluctant to give a major role to Darwinian competition at the precellular level, when we do not need it and we are much better off without it. Later, of course, things changed and natural selection became much more important, but this does not mean that it ever became the exclusive mechanism of evolution.

It has been shown recently that there have been at least two other major means of modification. One is the non-adaptive mode described by Motoo Kimura in 'The Neutral Theory of Molecular Evolution' (1979 and 1984); the other was proposed by Gabriel Dover in a paper entitled 'Molecular drive: a cohesive mode of species evolution' (1982).

At present it is too early to establish the relative importance of the various modes, but there is little doubt that the conclusions of Kimura and Dover are strongly supported by the evidence from molecular biology. Even without going into further details, therefore, we obtain the same conclusion as the previous section: non-adaptive processes do exist in Nature, and did play a major role in the history of life.

THE OTHER MECHANISM

In the Introduction to the Origin of Species, Darwin wrote: "I am convinced that Natural Selection has been the main but not exclusive means of modification". Wallace, for his part, went so far in emphasizing the insufficiency of selection (particularly in the evolution of the mind), that in 1870 Darwin wrote to him: "I hope that you have not completely killed Your and mine creature". The search for 'another' mechanism of evolution started in this way, and has gone on ever since.

D'Arcy Thompson, Alfred Whitehead, Conrad Waddington, Paul Weiss, Walter Elsasser, Ludwig von Bertalanffy, Karl Popper and many others have emphasized the need to go beyond natural selection with arguments from all natural sciences.

Furthermore, the 'Catastrophe Theory' of René Thom (1972), and the 'Thermodynamics of Evolution' of Ilya Prigogine (1972), have given us — from very different viewpoints — algorithms, equations and concepts for entirely new approaches to the description of Nature.

On the other hand, the Modern Synthesis of the 1930s and 40s used natural selection as the only effective mechanism of evolution, and many biologists accept this. Even the evidence from molecular biology and the rigorous conclusions of scientists like Kimura and Dover do not seem to have changed the belief that natural selection is the all-important source of biological modification.

I believe that there is a profound meaning in this, and that biologists have not simply been stubborn. On the contrary, I think that natural selection has ruled for two very logical reasons.

One is that we cannot speak of *two* mechanisms of evolution with *one* concept of adaptation because of the deep belief that adaptation is the heart and soul of evolution. The second reason is the classical theory of the cell: as we have seen before, if the cell is a duality of genotype and phenotype, the mechanism of change is bound to be a two-step process of variation in the genotypes and selection in the phenotypes.

In this book, on the other hand, we have seen that we do have a new concept of adaptation as well as a new theory of the cell, and we can legitimately say therefore that we do have a new theoretical framework for evolution.

The experiments in population genetics which study the mechanism of evolution, are set up *a priori* as experiments of adaptations of organisms to niches, and in this way they are bound to be tests for natural selection only. They are perfectly valid experiments, of course, because they truly test if natural selection works or not, but that is all they can tell us. This is what the concept of 'adaptation to the environment' really means. It is a theoretical telescope that makes us look at a very real dimension of Nature, but if we only look through that we cannot see any other dimension, and therefore no other mechanism seems to exist.

But let us try another telescope. To the adaptation of organisms to niches, let us add the adaptation of organisms and niches to natural cycles, and we see a very different panorama.

We realize that organisms adapt to niches by natural selection, while organisms and niches together adapt to the cycles of life by natural conventions. With two distinct concepts of adaptation we have immediately two distinct mechanisms of evolution. Notice also that the new mechanism does not work on mysterious phenomena that are presently inaccessible. The natural cycles are all around us, and we are well aware of their importance.

The panorama that we have before us is therefore a new theory: biological evolution took place by natural conventions and by natural selection, as linguistic evolution took place by linguistic conventions and by linguistic selection. This is why I call it 'the Semantic theory of evolution'.

THE OLD AND THE NEW HYPOTHESES

The analogy between organic and linguistic evolution has been suggested by various authors in our century, but has not been taken seriously as an alternative to the classical theory. In my opinion, this is because there has been a major discrepancy between linguistics and biology. In linguistics, people have always been aware that a language is not really a duality of langue and parole because there is an essential third party between them. That party, of course, is the brain. Even if we study a language at a purely abstract level, we cannot forget that fundamental intermediary because the creation of linguistic meanings cannot be attributed to either langue or parole, but only to the brain.

In biology, on the contrary, the duality of genotype and phenotype has been genuinely regarded as a true scheme, and the importance of the ribotype has been systematically played down. The parallel between organisms and languages, in other words, had no deep meaning because the role of the ribotype in biology was nothing like the role of the brain in linguistics. And yet the ribotype is truly the brain of the cell, since it does for the cell what the brain does for a language: it creates meaning. All the components of the genetic code belongs to the ribotype, and it is the code which creates biological meaning.

We realize in this way that a true semantic approach to the history of life was not possible without the ribotype concepts.

The Semantic theory of evolution depends upon the Ribotype theory of the cell because it needs the description of the cell as a trinity of fundamental categories. It also depends upon the Ribotype theory of the origins, because the idea of adaptation to natural cycles requires a consistent reconstruction of the cycle which first came into existence and which is the basis of all other cycles of life: the cell cycle.

Let us summarize. The Classical (or Synthetic) theory of evolution was based on three hypotheses:

1) The cell is a duality of genotype and phenotype.

2) Biological adaptation is the adaptation of organisms to the environment.

3) The primary mechanism of modification is evolution by natural selection.

The Semantic theory is based on the following hypotheses:

1) The cell is a trinity of genotype, ribotype and phenotype.

2) Biological adaptation is adaptation to natural cycles in the first place and adaptation to the environment in the second place.

	The Classical theory	The Semantic theory
Nature of the Cell	Genotype Phenotype	Genotype Ribotype Phenotype
Types of Adaptation	Adaptation to the environment	Adaptation to natural cycles Adaptation to the environment
Mechanism of Evolution	Natural Selection	Natural Conventions Natural Selection

FIGURE 14:   Basic principles of the Classical and the Semantic theory.

3) The two primary mechanisms of modification are evolution by natural conventions in the first place, and evolution by natural selection in the second place.

The scheme of Figure 14 illustrates the difference between the two theories and shows that there is both continuity and discontinuity between them. There is continuity because all the concepts of the Classical theory continue to appear in the new framework. There is also discontinuity, however, because the main role is shifted to the new concepts.

CONCLUSION

Linneas wrote the 'Systema Naturae', in 1735, not just as a description but as an interpretation of Nature. The Systema was to be the scientific equivalent of the Great Chain of Being of Aristotle, of the idea that the world is an orderly procession of forms from minerals to vegetables, from vegetables to animals, and from animals to intelligent beings.

In this hierarchy every species is a closed world, but is also linked to all other species because together they form the Great Ladder of Life where every creature has a place and a purpose. For Aristotle and Linnaeus this ideal structure, this system of worlds within higher worlds, does not change: species are forever, and life was given complete at the beginning.

It may seem a paradox, but from the point of view of evolution the Systema Naturae was an ideal starting point, because it was enough to introduce history into it for the Great Ladder to become the Tree of Life. The Chain of Being was thus transformed into the Chain of Becoming, but in the process something has been lost. The hierarchy, the eternal values, the sense of purpose and of belonging, have all gone. They have been replaced by a continuous change towards nowhere, fueled by competition and selfishness.

D'Arcy Thompson had an intense dislike for the theory of Darwin where Nature seems to be imposing taxes on every creature, to demand a profit for the right to exist, to nourish and favour only ruthless exploitation.

By the theory of natural selection, every variety of form and colour was urgently and absolutely called upon to produce its title to existence either as an active useful agent, or as survival.

His protest is that there is beauty for its own sake in Nature, there is harmony without profit, there are timeless laws of mathematics and physics to respect far above the fiscal law of selection.

Cell and tissue, shell and bone, leaf and flower, are so many portions of matter, and it is in obedience to the laws of physics that their particles have been moved, moulded and conformed.

D'Arcy Thompson sought and found refuge in the ideal world of the Classics.

The perfection of mathematical beauty is such that whatsoever is most beautiful and regular is also found to be most useful and excellent. Not only the movements of the heavenly host must be determined by observation and elucidated by mathematics, but whatsoever else can be expressed by number and defined by natural law. This is the teaching of Plato and Pythagoras, and the message of Greek wisdom to mankind.

The idea of the natural cycles brings back something of those lost values. The cycles are stable, are self-perpetuating, and form a hierarchy of cycles within higher cycles. Above all, they bring back the idea that all creatures are interdependent, an idea that was shared by many primitive cultures.

Biologists have always been aware of it, but in the Darwinian framework where selection and competition are the supreme rule, the idea becomes barely believable. This is what shocked and divided scientists. Natural selection is real, historical change is real, and they do not go against the laws of mathematics and physics, and if they show that Nature has an ugly face, why cover it up? Why dream of a world of beauty and perfection and not look straight into the eyes of the real world? The childhood of culture is over, and if natural selection is the supreme law of life, so be it.

But D'Arcy Thompson was right: natural selection is not above everything. On this point the Semantic theory is in sharp contrast with Darwinism and with the Modern Synthesis. The supreme law of life is cooperation, not competition. Competition is real, but adaptation to natural cycles is fundamental. First and foremost the forms of life had to create natural cycles by co-operating and complementing one another. Only after that was there some space for competition and minor readjustments. Natural selection is the mechanism of these readjustments, and necessarily takes second place.

One other point must be emphasized. The Semantic theory gives a secondary importance to competition, but what is put in its place is not just another rigid law, because the natural conventions are a very heterogeneous family. There are major and minor conventions in Nature as there are major and minor grammatical rules in a language.

The world of Darwin was dominated by the monolithic law of universal competition, while the world of the Semantic theory is made of a multitude of biological conventions which form a unity just as the various conventions of grammar and syntax form the pluralistic and evolving unity of a language.

There are two main differences between organisms and languages. The first is that various languages evolved independently on earth, and the parallel with life would be complete only if we could compare the creatures that evolved independently on different planets. This would really make us appreciate the difference between natural conventions and natural selection as easily as we differentiate between linguistic conventions and linguistic selection.

The second difference is that information and the carrier of information are separate in a language, while in life they are one. When the nucleus sends a messenger-RNA to the cytoplasm, the messenger 'is' the message. It is as if Nature writes the word 'apple', for example, and then the word folds itself up and becomes a real apple. This deep unity of structure and function gives a material reality to life that does not exist in the abstract world of a language, and makes it difficult for us to realize not just that life *has* a language, but the more subtle idea that life *is* a language.

From this point of view, the Semantic theory can be summarized by saying that *life is the language that Nature learned to speak on our planet.* There is meaning in life as there is meaning in every language, and perhaps we will find it if we listen carefully.

We may discover that Nature is trying to tell us what T.S. Eliot said when he wrote:

> the end of all our exploring
> will be to arrive where we started
> and to know the place for the first time.

References

Ager, D.V., 1973, *The Nature of Stratigraphic Record*, London, MacMillan.

Allen, G.E., 1975, *Life Science in the Twentieth Century*, New York, John Wiley.

Altmann, K., 1890, *Die Elementarorganismen und ihre Beziehungen zu den Zellen*, Leipzig, Viet.

Anfinsen, C.B. and Haber, E., 1961, Studies on the reduction and reformation of protein disulfide bonds. *J. Biol. Chem.*, **236**, p. 1361-1363.

Aristotle, (originally published ca. 350 BC), 1910, *The Reproduction of Animals (De Generatione Animalium)*, English translation by A. Platt, Oxford, Clarendon Press.

Arrhenius, S., 1908, *Worlds in the Making. The Evolution of the Universe*, New York, Harper and Row.

Avery, O.T., Macleod, C.M. and McCarty, M., 1944, Studies on the chemical nature of the substance inducing transformation of pneumococcal types. Induction of transformation by a desoxyribonucleic acid fraction isolated from pneumococcus Type III. *J. exp. Med.*, **79**, p. 137-158.

Awramik, S.M., 1980, The pre-Phanerozoic biosphere: three billion years of crises and opportunities. In *Biotic Crises in Ecology and Evolutionary Time*, ed. by M.H. Nitecki, New York, Academic Press.

Baltimore, D., 1970, RNA-dependent DNA polymerase in virions of RNA tumour viruses. *Nature*, **226**, p. 1209-1211.

Barbieri, M., 1974, The primitive ribosome model. *J. Theor. Biol.*, **47**, p. 269-280.

Barbieri, M., 1979, Ribosome crystallization in homogenates and cell extracts of chick embryos. *J. Supramol. Structure.* **10**, p. 349-357.

Barbieri, M., 1981, The Ribotype Theory on the Origin of Life. *J. Theor. Biol.*, **91**, p. 545-601.

Barbieri, M., 1983, Ribosome crystallization in chick embryos: facts and theories. *Rivista di Biologia*, **76**, p. 65-95.

Barbieri, M., 1983, The Preribosome Reticulum. *Rivista di Biologia*, **76**, p. 409-434.

Barbieri, M., 1983, The Preribosome Theory on Mitosis and Differentiation. *Rivista di Biologia*, **76**, p. 639-653.

Barbieri, M., Pettazzoni, P., Bersani, F. and Maraldi, N.M., 1970, Isolation of ribosome microcrystals. *J. Mol. Biol.*, **54**, p. 121-124.

Barghoorn, E.S., 1971, The oldest fossils. *Scientific American*, **224**, p. 30-42.

Bernal, J.D., 1951, *The Physical Basis of Life*, London, Routledge and Kegan Paul.

Bernal, J.D., 1967, *The Origin of Life*, London, Weidenfeld and Nicholson.

Bertalanffy, L. von, 1968, *General System Theory. Foundation, Development, Applications*, New York, Braziller.

Blith, E., 1835, An attempt to classify the Variety of Animals. *Magazine of Natural History*, 8, p. 40-59.

Boivin, A. and Vendrely, R., 1947, Sur le rôle possible deux acides nucleiques dans la cellule vivante. *Experientia*, 3, p. 32-34.

Brachet, J., 1941, La localisation des acides pentosennucleiques dans les tissues animaux et les oeufs d'amphibiens en voie de dévelopment. *Archives de Biologie*, 53, p. 207-257.

Brachet, J., 1946, Nucleic acids in the cell and the embryo. *Symp. Soc. Exp. Biol.*, 1, p. 213-215, 222.

Brachet, J., 1960, *The Biological role of the ribonucleic acids*, Amsterdam, Elsevier.

Britten, R.J. and Kohne, D.E., 1968, Repeated sequences in DNA. *Science*, 161, p. 529-540.

Britten, R.J. and Davidson, E.H., 1969, Gene regulation for higher cells: a theory. *Science*, 165, p. 349-357.

Britten, R.J. and Davidson, E.H., 1971, Repetitive and non-repetitive DNA sequences and a speculation on the origin of evolutionary novelty. *Quarterly Review of Biology*, 46, p. 111-138.

Broda, E., 1975, *The Evolution of the Bioenergetic Processes*, Oxford, Pergamon Press.

Broda, E. and Peschek, G.A., 1979, Did respiration of photosynthesis come first? *J. Theor. Biol.*, 81, p. 201-212.

Bruno, G., Originally published in 1584, 1980, *De l'infinito, universo e mondi*. English translation in *The Quest for Extraterrestrial Life*, ed. by D. Goldsmith, California, University Science Books.

Busch, H. and Smetana, K., 1970, *The Nucleolus*, New York and London, Academic Press.

Byers, B., 1966, Ribosome crystallization induced in chick embryo tissues by hypothermia. *J. Cell Biol.*, 30, p. C1-C6.

Cairns-Smith, A.G., 1966, The origin of life and the nature of the primitive gene. *J. Theor. Biol.*, 10, p. 53-88.

Cairns-Smith, A.G., 1982, *Genetic Takeover and the mineral origins of life*, Cambridge, Cambridge University Press.

Calvin, M., 1969, *Chemical Evolution*, Oxford, Clarendon Press.

Carroll, L., 1865, *Through the Looking-Glass*, London, MacMillan.

Caspersson, T., 1941, Studien uber die Eiweisseumsatz der Zelle. *Die Naturwissenschaften*, 29, p. 33-43.

Cech, T.R., Zaug, A.J. and Grabowski, P.J., 1981, In vitro splicing of the ribosomal RNA precursor of Tetrahymena: involvement of a guanosine nucleotide in the excision of the intervening sequence. *Cell*, 27, p. 487-496.

Cech, T.R., Zaug, A.J., Grabowski, P.J. and Brehm, S.L., 1982, Transcription and splicing of the ribosomal RNA precursor of Tetrahymena. *The Cell Nucleus*, 10, p. 171-204, New York, Academic Press.

Ceruti, A., 1942, Il rapporto acido nucleinico-proteina. *Archivio Botanico*, 18, p. 5-34.

Chomsky, N., 1965, *Aspects of the Theory of Syntax*, Cambridge, Mass., MIT Press.

Chomsky, N., 1972, *Problems of Knowledge and Freedom*, London, Fontana.

Claude, A., 1940, Particulate components of normal and tumour cells. *Science*, 91, p. 77-78.

Constantin, J., 1923, *Origine de la Vie sur le Globe*, Paris.

Cook, N.D., 1977, The case for Reverse Translation. *J. Theor. Biol.*, 64, p. 113-135.

Crick, F.H.C., 1957, The structure of nucleic acids and their role in protein synthesis. *Biochem. Soc. Symp*, 14, p. 25-26.

Crick, F.H.C., 1958, On Protein Synthesis. *Symp. Soc. Exp. Biol.*, 12, p. 155.

Crick, F.H.C., 1970, Central Dogma of molecular biology. *Nature*, 227, p. 561-562.

Crick, F.H.C., 1981, *Life Itself, its Origin and Nature*, New York, Simon & Schuster.

Cuvier, G., 1805, *Lecons d'Anatomie Comparee*, Paris.

Cuvier, G., 1812, *Recherches sur les ossements fossiles*, Paris.

Darnell, J.E., 1978, Implications of RNA splicing in Evolution of Eukaryotic Cells. *Science*, 202, p. 1257-1260.

Darwin, C.R., 1859, *On the Origin of Species by Means of Natural Selection*, London, John Murray.

Darwin, C.R., 1868, *The variation of animals and plants under domestication*, London, John Murray.

Darwin, C.R. and Wallace, A.R., 1858, On the Tendency of Species to form Varieties; and on the perpetuation of Varieties and Species by Natural Means of Selection. *J. of the Proceedings of the Linnean Society*, 3, p. 62.

Dawkins, R., 1976, *The Selfish Gene*, Oxford, Oxford University Press.

D'Herelle, F.H., 1917, Sur un microbe invisible antagoniste des bacilles dysenteriques. *Comptes Rendus Acad. Sci.*, Paris, 165, p. 373-374

Dobzhansky, T., 1937, *Genetics and the Origin of Species*, New York, Columbia University Press.

Doolittle, W.F., 1978, Genes in pieces: were they ever together? *Nature*, 272, p. 581-582.

Doolittle, W.F. and Sapienza, C., 1980, Selfish genes, the phenotype paradigm and genome evolution. *Nature*, 284, p. 601-603.

Dougherty, E.C., 1957, Neologisms needed for structures of primitive organisms. *J. Protozoology*, 4 (Suppl.), p. 14.

Dounce, A.L., 1952, Duplicating mechanism for peptide chain and nucleic acid synthesis. *Enzymologia*, **15**, p. 251-258.

Dover, G., 1982, Molecular drive: a cohesive mode of species evolution. *Nature*, **299**, p. 111-117.

Drummond, R., 1980, The critical role of the nucleolus in cell differentiation and stem cell development with particular reference to its importance in imaginal development, spermatogenesis and haemopoiesis — A new fundamental concept. *Medical Hypotheses*, **6**, p. 1221-1247.

Drummond, R., 1980, The critical role of the nucleolus in cell differentiation and stem cell development — An extension of the concept with a look at the mast cell and its possible role in zinc metabolism. *Medical Hypotheses*, **6**, p. 1249-1274.

Eigen, M. and Schuster, P., 1977, The Hypercycle. A Principle of Natural Self-Organization. *Die Naturwissenschaften*, **64**, p. 541-565.

Eiseley, L.C., 1959, Charles Darwin, Edward Blith, and the theory of natural selection. *Proc. Amer. Phil. Soc.*, **103**, p. 94-158.

Eldredge, N. and Gould, S.J., 1972, Punctuated equilibria: an alternative to phyletic gradualism. In *Models in Paleobiology*, ed. by T.J.M. Schopf, p. 82-115, San Francisco, Freeman, Cooper & Co.

Elsasser, W.M., 1958, *The Physical Foundation of Biology*, New York and London, Pergamon Press.

Elsasser, W.M., 1981, Principles of a New Biological Theory: a summary. *J. Theor. Biol.*, **89**, p. 131-150.

Fischer, E., 1906, *Untersuchungen über Aminosäuren, Polypeptide and Proteine*, Berlin, Springer Verlag.

Fischer, E., 1907, *Untersuchungen in der Puringruppe*, Berlin, Springer Verlag.

Fisher, R.A., 1930, *The Genetical Theory of Natural Selection*, Oxford, Clarendon Press.

Fontana, 1774, Quotation from Montgomery T.H. (1898) 'Comparative cytological studies with special regard to the nucleolus'. *J. Morphol.*, **15**, p. 265-564.

Fox, S.W. and Dose, K., 1972, *Molecular Evolution and the Origin of Life*, San Francisco, Freeman.

Fox, S.W., 1978, The Origin and Nature of Protolife. In *The Nature of Life*, ed. by W.H. Heidecamp, p. 23-92, Baltimore, University Park Press.

Gamow, G., 1954, Possible relation between deoxyribonucleic acid and protein structures. *Nature*, **173**, p. 318.

Gatlin, L.L., 1972, *Information Theory and the Living System*, New York, Columbia University Press.

Gatlin, L.L. 1977, Meaningful Information Creation: an alternative interpretation of the Psi Phenomenon. *J. Amer. Soc. for Psychical Research*, **71**, p. 1-18.

Ghiara, G. and Taddei, C., 1966, Dati citologici e ultrastrutturali su di un particolare tipo di costituenti basofili del citoplasma di cellule follicolari e di ovociti ovarici di rettili. *Boll. Soc. Ital. Biol. Sper.*, **42**, p. 784-788.

Gilbert, W., 1978, Why genes in pieces? *Nature*, **271**, p. 501.

Goldschmidt, R., 1940, *The Material Basis of Evolution*, New Haven, Yale University Press.

Gould, S.J., 1978, *Ever Since Darwin*, New York, Burnett Books.

Gould, S.J., 1981, The chance that shapes our ends, *New Scientist*, **89**, p. 347-349.

Gould, S.Y., Raup, D.M., Sepkoski, J.J., Schopf, T.J.M. and Simberloff, D.S., 1977, The shape of evolution: a comparison of real and random clades. *Paleobiology*, **3**, p. 23-40.

Griffith, F., 1928, Significance of pneumococcal types. *J. Hygiene*, **27**, p. 113-159.

Haeckel, E., 1866, *Generalle Morphologie der Organismen*, Berlin, Georg Reimer.

Haeckel, E., 1878, *Das Protistenreich*, Leipzig, Gunther.

Haldane, J.B.S., 1929, The Origin of Life. *Rationalist Annual*, **3**, p. 148-153.

Haldane, J.B.S., 1932, *The causes of Evolution*, New York, Harper.

Haller, A. von, 1744, Hermanni Boerhaave praelectiones academicae. In *Praelectiones academicae in proprias institutiones rei medicae edidit, et notas addidit Albertus Haller* by Hermann Boerhaave, Gottingen, A. Vandenhoeck.

Hamilton, W.D., 1964, The genetical evolution of social behaviour. *J. Theor. Biol.*, **7**, p. 1-16 and p. 17-32.

Henderson, L.J., 1913, *The Fitness of the Environment*, New York, Macmillan.

Henderson-Sellers A. and Cogley, J.G., 1982, The Earth's early hydrosphere. *Nature*, **298**, p. 831-835.

Hippocrates (originally published ca. 400 BC), 1939, *On airs, waters and places*, English translation by F. Adams, London, Bailliere, Tindall and Cox.

Hoagland, M.B., Zamecnik, P.C. and Stephenson, M.L., 1957, Intermediate reactions in protein biosynthesis. *Biochem. Biophys. Acta*, **24**, p. 215-216.

Horrobin, D., 1981, Nucleus or nucleolus: which runs the cell? *New Scientist*, **89**, p. 266-269.

Hoyle, F. and Wickramasinghe, C., 1978, *Lifeclouds*, London, Dent.

Hoyle, F. and Wickramasinghe, C., 1979, *Diseases from Space*, London, Dent.

Hoyle, F. and Wickramasinghe, C., 1981, *Evolution from Space*, London, Dent.

Huxley, J.S., 1942, *Evolution, The Modern Synthesis*, London, Allen and Unwin.

Jacob, F. and Monod, J., 1961, Genetic regulatory mechanisms in the synthesis of proteins. *J. Mol. Biol.*, **3**, p. 318-356.

Johannsen, W., 1909, *Elemente der exakten Erblichkeitslehre*, Jena, Gustaf Fisher.

Kaltschmidt, E. and Wittmann, H.G., 1970, Two-dimensional polyacrylamide gel electrophoresis for fingerprinting of ribosomal proteins. *Anal. Biochem.*, **35**, p. 401-412.

Kendrew, J.C. et al., 1958, A three-dimensional model of the myoglobin molecule obtained by X-ray analysis. *Nature*, **181**, p. 662-666.

Kimura, M., 1979, The Neutral Theory of Molecular Evolution. *Scientific American*, **241**, No. 5, p. 94-104.

Kimura, M., 1983, *The Neutral Theory of Molecular Evolution*, Cambridge, Cambridge University Press.

Kuckuck, M., 1907, *Die Lösung des Problems der Urzeugung*, Leipzig.

Lamarck, J.P.B.A. de M., (originally published in 1809), 1914, *Philosophie Zoologique*, English translation by H. Elliot, London, MacMillan. Reprinted 1963, New York, Hafner.

Le Conte, J., 1888, Evolution and its Relation to Religious Thought. Quoted by A.O. Lovejoy in *Forerunners of Darwin*, p. 379.

Lederberg, J., 1952, Cell genetics and hereditary symbiosis. *Physiol. Review*, **32**, p. 403-430.

Le Duc, S., 1907., *Les bases physiques de la vie*, Paris.

Leeuwenhoek, A. van, 1677, Philosophical transactions of the Royal Society 13th year, No. 133, p. 821-831.

Lewontin, R.C., 1978, Adaptation, *Scientific American*, **239**, No. 3, p. 157-169.

Linnaeus, C., 1735, *Systema naturae, sive regna tria naturae systematice proposita per classes, ordines, genera et species*, Lugduni Batavorum, Lyon, Theodorum Haak.

Lovejoy, A.O., (originally published in 1909), 1968, The Argument for Organic Evolution before the Origin of Species. In *Forerunners of Darwin*, ed. by B. Glass, p. 356-414, Baltimore, The John Hopkins Press.

Løvtrup, S., 1977, *The Phylogeny of Vertebrata*, London, John Wiley.

Lucretius (Titus Lucretius Carus, ca. 55 BC), 1951, *De Rerum Natura*, English translation by R.E. Latham, Harmondsworth, England, Penguin Books.

Luria, S.E. and Delbruck, M., 1943, Mutation of bacteria from virus sensitivity to virus resistance. *Genetics*, **28**, p. 491-511.

Margulis, L., 1981, *Symbiosis in Cell Evolution*, San Francisco, Freeman.

Mayr, E., 1942, *Systematics and the Origin of Species*, New York, Columbia University Press.

Mayr, E., 1978, Evolution, *Scientific American*, **239**, No. 3, p. 39-47.

Mekler, L.B., 1967, Mechanism of Biological Memory. *Nature*, **215**, p. 481-484.

Mendel, G., 1866, Versuche über Pflanzen-Hybriden. *Verh. Naturforsch. Ver. Brunn*, **4**, p. 3-47.

Mereschkovsky, K.C., 1905, Le plant considéré comme un complex symbiotique. *Bull. Soc. Nat. Sci. Ouest*, **6**, p. 17-98.

Miescher, F., 1871, Ueber die chemische Zusammensetzung der Eiterzellen. In Hoppe-Seyler's *Medicinisch-chemische Unterschungen* p. 441-460, Berlin, August Hirschwald.

Miller, S.L., 1953, A production of amino acids under possible primitive earth conditions. *Science*, **117**, p. 528-529.

Miller, S.L. and Orgel, L.E., 1974, *The Origins of Life on the Earth*, Englewood Cliffs, New Jersey, Prentice-Hall.

Monod, J. and Jacob, F., 1962, General conclusion: teleonomic mechanisms in cellular metabolism, growth and differentiation. *Cold Spring Harbor Symp.*

Quant. Biol., **26**, p. 389-401.

Monod, J., Changeaux, J.P. and Jacob, F., 1963, Allosteric proteins and cellular control systems. *J. Mol. Biol.*, **6**, p. 306.

Monod, J., 1970, *Le Hasard et la Necéssité*, Paris, Editions du Seuil. Published in English in 1971, *Chance and Necessity*, New York, Knopf.

Ninio, J., 1979, *Approaches Moleculaires de l'Evolution*, Paris, Masson. Published in English in 1982 as *Molecular Approaches to Evolution*, London, Pitman.

Niremberg, M.W. and Matthaei, J.H., 1961, The dependence of cell-free protein synthesis in E.coli upon naturally occurring or synthetic polyribonucleotides. *Proc. Nat. Acad. Sci. USA*, **47**, p. 1588-1602.

Oparin, A.I., 1924, *Proiskhozhdenie Zhizni*, Moscow, Moskovskii Rabochii. English translation in 1967, *The Origin of Life*, ed. by Bernal, p. 199-234, London, Weidenfeld and Nicholson.

Oparin, A.I., 1936, *Vozniknovie Zhizni na Zemle*, Moscow, Biomedgiz. English translation by S. Morgulis, 1938, in *The Origin of Life*, New York, Macmillan. Reprinted 1953, New York, Dover.

Oparin, A.I., 1957, *The Origin of Life on the Earth*, Edinburgh, Oliver and Boyd.

Oparin, A.I., 1968, *Genesis and evolutionary development of Life*, New York, Academic Press.

Orgel, L.E. and Lohrmann, R., 1974, Prebiotic chemistry and nucleic acid replication. *Accounts of Chemical Research*, **7**, p. 368-377.

Orgel, L.E. and Crick, F.H.C., 1980, Selfish DNA: the ultimate parasite. *Nature*, **284**, p. 604-607.

Osborn, H.F., 1918, *The Origin and evolution of life*, London.

Ospovat, D., 1981, *The Development of Darwin's Theory*, Cambridge, Cambridge University Press.

Pantin, C.F.A., 1951, Organic Design. *Advancement of Science*, **8**, p. 138-150.

Pasteur, L. 1860, Expériences relatives aux générations dites spontanées. *Comptes Rendus Acad. Sci. Paris*, **50**, p. 303.

Pasteur, L., 1860, De l'origine des ferments. Nouvelles expériences relatives aux générations dites spontanees. *Comptes Rendus Acad. Sci. paris*, **50**, p. 849.

Pauling, L., Corey, R.B. and Branson, H.R., 1951, The structure of proteins: two hydrogen-bonded helical configurations of the polypeptide chain. *Proc. Nat. Acad. Sci. USA*, **37**, p. 205-211.

Perrier, E., 1920, *La terre avant l'histoire. Les origines de la vie et de l'homme*, Paris.

Perutz, M.F., et al., 1960, Structure of Haemoglobin: a three-dimensional Fourier synthesis at 5A resolution obtained by X-ray analysis. *Nature*, **185**, p. 416-422.

Ponnamperuma, C., 1972, *Origins of Life*, London, Thames and Hudson.

Popper, K.R., 1934, *Logik der Forschung*, Vienna, Springer Verlag. Published in English in 1959 as *The Logic of Scientific Discovery*, London, Hutchinson & Co.

Popper, K.R., 1983, *Realism and the Aim of Science*, Volume I of *The Postscript*, London, Hutchinson & Co.

Portier, P., 1918, *Les Symbiotes*, Paris, Masson et Cie.

Prigogine, I., 1972, Thermodynamics of Evolution. *Physics Today*, **25**, p. 2.

Racker, E., 1976, *A New Look at Mechanisms in Bioenergetics*, New York, Academic Press.

Raup, D.M., 1979, Conflicts between Darwin and Paleontology. *Bulletin, Chicago Field Museum of Natural History*, **50**.

Redi, F., 1668, *Esperienze intorno alla generazione degli insetti*, Firenze.

Roberts, R.B., 1958, *Microsomal Particles and Protein Synthesis*, Washington, Pergamon Press.

Sagan, L., 1967, On the origin of mitosing cells. *J. Theor. Biol.*, **14**, p. 225-275.

Sanger, F. and Thompson, E.O.P., 1953, The amino-acid sequence in the glycyl chain of Insulin. *Biochemical Journal*, **53**, p. 366-374.

Saussure, F. de, 1916, *Cours de linguistique generale*, Paris, Payot.

Schimper, A.F.W., 1883, Uber die Entwickelung der Chlorophyllkörner und Farbkorper. *Bot. Ztg.*, **41**, p. 105-114.

Schleiden, M., 1842, *Grundzüge der Wissenschaftlichen Botanik*, Leipzig, W. Engelmann.

Schopf, J.W., 1978, The evolution of the earliest cells. *Scientific American*, **239**, No. 3, p. 84-103.

Schwann, T., 1839, *Microscopische Untersuchengen über die Uebereinstimmung in der Strukter und dem Wachsthum der Thiere und Pflanzen*, Berlin, Reimer.

Sepkoski, J.J., 1978, A Kinetic Model of Phanerozoic Taxonomic Diversity. *Paleobiology*, **4**, p. 223-251.

Sermonti, G., 1981, Dibattito sull'Evoluzionismo — Debate on Evolutionism. *Rivista di Biologia*, **74**, p. 359-400.

Simpson, G.G., 1944, *Tempo and Mode in Evolution*, New York, Columbia University Press.

Simpson, G.G., 1949, *The Meaning of Evolution*, New Haven, Yale University Press.

Spallanzani, L., 1795, *Saggio di osservazioni microscopiche concernenti il sistema della generazione de' signori di Needham e Buffon*, Modena.

Spiegelman, S., et al., 1970, Characterization of the products of RNA-directed DNA polymerases in oncogenic RNA viruses. *Nature*, **227**, p. 563-567.

Stanier, R.Y., 1970, Some aspects of the biology of cells and their possible evolutionary significance. In *Organization and Control in Prokaryotic and Eukaryotic Cells*, ed. by H.P. Charles and B.C. Knight, p. 1-38, Cambridge, Cambridge University Press.

Stanier, R.Y., Doudoroff, M. and Adelberg, E.A., 1963, *The Microbial World*, Englewood Cliffs, New Jersey, Prentice-Hall.

Stanley, S.M., 1975, A theory of evolution above the species level. *Proc. Nat. Acad. Sci. USA*, **72**, p. 650.

Stanley, S.M., 1979, *Macroevolution*, San Francisco, Freeman.

Steele, E.J., 1979, *Somatic Selection and Adaptive Evolution (On the Inheritance of Acquired Characters)* Toronto, Williams-Wallace.

Temin, H.M., 1963, The effects of Actinomycin-D on growth of Rous sarcoma virus. *Virology*, **20**, p. 577-582.

Temin, H.M. and Mizutani, S., 1970, RNA-dependent DNA polymerase in virions of Rous sarcoma virus. *Nature*, **226**, p. 1211-1213.

Thom, R., 1972, *Stabilité structurelle et morphogénèse.* Published in English in 1975 as *Structural Stability and Morphogenesis,* Reading, Massachusetts, W.A. Benjamin.

Thompson D'Arcy, W., 1917, *On Growth and Form,* Cambridge, Cambridge University Press.

Tyler, S.A. and Barghoorn, E.S., 1954, Occurrence of structurally preserved plants in Precambrian rocks of the Canadian shield. *Science*, **119**, p. 606-608.

Urey, H.C., 1952, On the early chemical history of the earth and the origins of life. *Proc. Nat. Acad. Sci. USA*, **38**, p. 351-363.

Valentine, J.W., 1978, The evolution of Multicellular Plants and Animals. *Scientific American*, **239**, No. 3, p. 104-117.

Van Valen, L., 1973, A new evolutionary law. *Evolutionary Theory*, **1**, p. 1-30.

Von Neumann, J., 1966, *Theory of Self-Reproducing Automata,* Urbana, University of Illinois Press.

Waddington, C.H., 1961, *The Nature of Life,* London, Allen and Unwin.

Waddington, C.H., 1969, *Towards a Theoretical Biology,* Chicago, Edinburgh University Press ad Aldine Press.

Wallace, A.R., 1855, On the Law which has regulated the Introduction of New Species. *The Annals and Magazine of Natural History*, **16/2**, p. 184-196.

Waller, J.P. and Harris, J.I., 1961, Studies on the composition of the proteins from Escherichia Coli ribosomes. *Proc. Nat. Acad. Sci. USA*, **47**, p. 18-23.

Wallin, J.E., 1927, *Symbionticism and the Origin of Species,* Baltimore, Williams and Wilkins.

Watson, J.D. and Crick, F.H.C., 1953, A structure for deoxyribose nucleic acid. *Nature*, **171**, p. 737-738.

Watson, J.D. and Crick, F.H.C., 1953, Genetical implications of the structure of deoxyribonucleic acid. *Nature*, **171**, p. 964-967.

Weiss, P., 1963, The Cell as Unit. *J. Theor. Biol.*, **5**, p. 389-397.

Weiss, P., 1969, The living system: determinism stratified. In *Beyond Reductionism*, ed. by A. Koestler and J.R. Smithies, p. 3-35, London, Hutchinson.

Whitehead, A.N., 1929, *Process and Reality,* Cambridge, Cambridge University Press.

Whittaker, R.H., 1959, On the broad classsification of organisms. *Quarterly Review of Biology*, **34**, p. 210-226.

Whittaker, R.H. and Margulis, L. (1978) Protist classification and the Kingdoms of Organisms. *BioSystems*, **10**, p. 3-18.

Wilson, E.B., 1900, *The Cell in Development and Inheritance,* New York, MacMillan.

Wilson, E.O., 1975, *Sociobiology: the New Synthesis,* Cambridge, Mass., Beknap.

Winogradski, S., 1889, Reserches physiologiques sur les sulphobacteries. *Ann. Inst. Pasteur*, 3, p. 49-60.

Winogradski, S., 1890, Reserches sur les organismes de la nitrification. *Ann. Inst. Pasteur*, 4, pp. 213-231, 257-275, 760-771.

Wittmann, H.G., Mussig, J., Piefke, J., Gewitz, H.S., Rheinberger, H.J. and Yonath, A., 1982, Crystallization of Escherichia Coli ribosomes. *FEBS Letters*, 146, p. 217-220.

Woese, C.R., 1970, Molecular Mechanism of Translation: a Reciprocating Ratchet Mechanism. *Nature*, 226, p. 817-820.

Woese, C.R., 1979, A proposal concerning the origin of life on the planet earth. *J. Mol. Evol.*, 13, p. 95.

Woese, C.R., 1980, Just So Stories and Rube Goldberg machines: speculations on the origin of the protein synthetic machinery. In *Ribosomes, Structure, Function and Genetics*, ed. by G. Chambliss *et al.*, p. 357-373, Baltimore, University Park Press.

Woese, C.R., 1981, Archaebacteria. *Scientific American*, 224, No. 6, p. 98-122.

Woese, C.R. and Fox, G.E., 1977, Phylogenetic structure of the prokaryotic domain: the primary kingdoms. *Proc. Nat. Acad. Sci. USA*, 74, p. 5088-5090.

Woese, C.R. and Fox, G.E., 1977, The concept of Cellular Evolution. *J. Mol. Evol.*, 10, p. 1-6.

Wool, I.G., 1980, The structure and function of eukaryotic ribosomes. In *Ribosomes, Structure, Function and Genetics*, ed. by G. Chambliss *et al.*, p. 797-824, Baltimore, University Park Press.

Wright, S., 1931, Evolution in Mendelian populations. *Genetics*, 16, p. 97-159.

Zuckerkandl, E., 1976, Programs of gene action and progressive evolution. In *Molecular Anthropology: Genes and Proteins in the Evolutionary Ascent of the Primates*, ed. by M. Goodman *et al.*, p. 387-447, New York, Plenum.

Zuckerkandl, E. and Pauling, L., 1965, Molecules as Documents of Evolutionary History. *J. Theor. Biol.*, 8, p. 357-366.

Author Index

Subject Index

185